Explaining Chemical Processes

Student Exercises and Teacher Guide for

Grade Ten Academic Science

Jim Ross — *The University of Western Ontario*

Mike Lattner — *Algonquin and Lakeshore Catholic District School Board*

rosslattner educational consultants — *London Ontario Canada*

| Authors | Jim Ross |
| | Mike Lattner |

| Printer | The Solski Group, Napanee, Ontario Canada |
| Cover Design | Images, London, Ontario Canada |

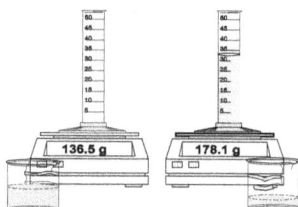

ISBN   978-1-897007-16-7

Offices   London Ontario Canada

To teachers, parents and students everywhere who desire to bring about new ways of understanding the world.

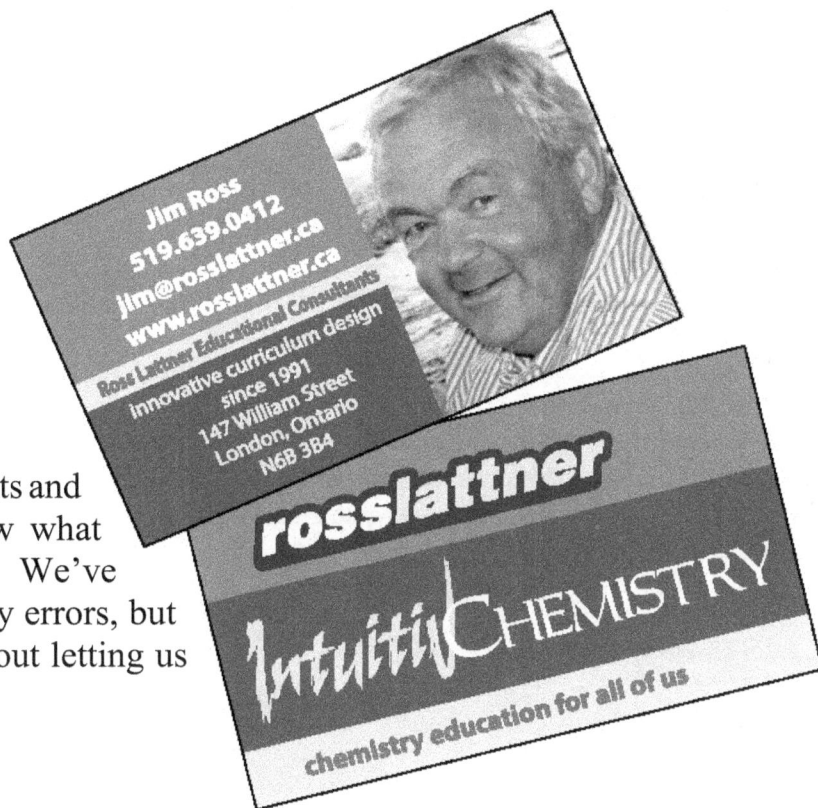

Jim Ross
519.639.0412
jim@rosslattner.ca
www.rosslattner.ca
Ross Lattner Educational Consultants
innovative curriculum design
since 1991
147 William Street
London, Ontario
N6B 3B4

rosslattner
Intuitiv CHEMISTRY
chemistry education for all of us

We welcome your comments and suggestions. Let us know what you find most useful. We've worked hard to remove any errors, but don't let a day go by without letting us know if you find one.

Stay in touch.

Our thanks to all of the wonderful people at the Faculty of Education, the University of Western Ontario.

Special thanks to Dick Bird, whose passion for chemistry started many students on their scientific explorations.

Explaining Chemical Processes

Table of Contents

Explaining Chemical Processes

Table of Contents

Explaining Chemical Processes

Teaching Chemical Processes

Title: Explaining Chemical Processes

Time Allocation: 27.5 hours (22 periods of 75 minutes each)

Authors: Jim Ross and Mike Lattner

Date: April 2004

Unit Description: An introduction to the simple chemical reactions, this unit begins with decomposition and synthesis reactions and moves to the reactions of acids and bases. The relationship between atomic structure, chemical behaviour, and organization of the periodic table is extended from the Grade Nine Chemistry unit. The unit is composed of three major sections.

1. Decomposition and synthesis reactions, with their one-to-many and many-to-one structure, are the easiest to work with. Within this context, students will learn how to represent chemical change four different, parallel ways: diagrams, symbolic balanced equations, word equations, and mass equations. We clearly intend this unit to anticipate and support the introduction of the mole concept in Grade 11.

2. Metals, non-metals and pH. The origins of acid and base behaviour are probed here.

3. Reactions of Acids and Bases. In both metal / acid reactions and acid / base reactions, we continue to use the four representations. Emphasis is placed on naming the chemicals involved.

At the end of each section is a thorough quiz.

Strand: Chemistry

Expectations: Overall Expectations: CPV 01 - 03
Specific Expectations: CP1.01 - .08; CH2.01 - .06

Explaining Chemical Processes

It is much easier to learn how to use a small number of dynamic explanatory propositions than to memorize a vast number of specific "facts".

Unit Planning Notes:

This unit proceeds from the previous units in Grade Nine Science. The educational challenge is to encourage the student to deal with chemistry from the most unifying point of view, that is, from the perspective of the atomic (ultimately electrical) constituents of matter. The alternative is to ask the student to deal with an infinity of instances. Therefore, this unit emphasizes the following:

Dalton's Theory of Chemical Change. This brief idea contains the foundational ideas. While it has been profoundly elaborated upon in the past two centuries, its essential points are still true and useful.

Arrhenius' theory of Acids and Bases. While it deals only with strong electrolytes in aqueous solutions, the conceptual clarity of this theory provides a powerful starting point for students.

Four representations of Chemical Change. Dalton's diagrams provide the beginning student with pictorial data that can support real insights. The more traditional balanced symbolic equation is far more abstract, but it has its foundations in the pictures. Word equations provide both the occasion and the support for learning chemical names. The mass equation supports later acquisition of the mole concept in Grade 11.

Prior Knowledge Required Students are expected to be familiar with the periodic table, with the basic concepts of chemical change, and with basic lab skills such as using the Bunsen burner, assembling apparatus, and measuring mass and volume.

You think there are "four classes of chemical reactions"? In which class, then, would you put the combustion of methane?

Teaching and Learning Strategies We wish to avoid the use of unnecessary categories or concepts. Accordingly, the "four classes of chemical reactions" are de-emphasized in favour of deeper, broader theoretical ideas with real explanatory power. We chose experiments whose outcomes can be either predicted or explained by resorting to these theories. Important ideas (the mass / mole concept, and nomenclature) are approached indirectly by using continuous exposure to mass equations and word equations.

Can a teacher evaluate understanding? Or do we evaluate student application, and infer understanding?

Assessment and Evaluation A variety of strategies are available. Day to day assessment of knowledge can follow the quizzes and the PEOE box diagrams. Clarity of communication can be assessed in the student's written explanations. Use the KICA (Knowledge, Inquiry, Communications, Applications wheel to indicate for students the learning area that you wish emphasize.

An average human
adult can keep track
of no more than 5 - 7
mental constructs at
one time. The
theories we teach
must respect the
limits of working
memory.

Dalton's Theory of Chemical Change First articulated two centuries ago, and encountered in Grade 9, this theory is still the foundation of chemistry.

1. All of the atoms present before a chemical change are still present after the change.
2. The total amount of mass (matter) is the same before and after a chemical change.
3. During a chemical change, atoms are rearranged into new groups.
4. During a chemical change, new particles of new substances are formed.
5. During a chemical change, old chemical bonds are broken, and new chemical bonds formed.
6. Chemical bonds are electrical forces of attraction between valence electrons and two nuclei.

2. **Arrhenius' Theory of Acids and Bases** explains many characteristics of acids and bases in water.

Students should use
these theoretical
propositions to explain
the phenomena they
encounter in this unit.

1. Acids are compounds which contain hydrogen ions (H^+).
2. Bases are compounds which contain hydroxide ions (OH^-).
3. Non-metal oxides form acidic solutions in water.
4. Metal oxides form basic solutions in water.
5. An acid plus a base react to produce neutral water, plus a salt.
6. A neutral solution contains equal numbers of H^+ and OH^- ions.

More general theories (Bronsted - Lowry, Lewis etc) explain more phenomena (weak acids and bases, non-aqueous solvents, etc) But using these theories requires distinctions that our students are not capable of making at this time. Better to use the simpler theories within their limits, and build upon that, than to introduce theories that our students cannot use.

The student who can
represent the chemical
changes encountered in
this manual in all 4
forms illustrated at
right, can be reasonably
thought to understand at
a descriptive level what
is happening during the
chemical change. A
dynamic explanation
(bonding, energy, rate
and equilibrium) must
wait.

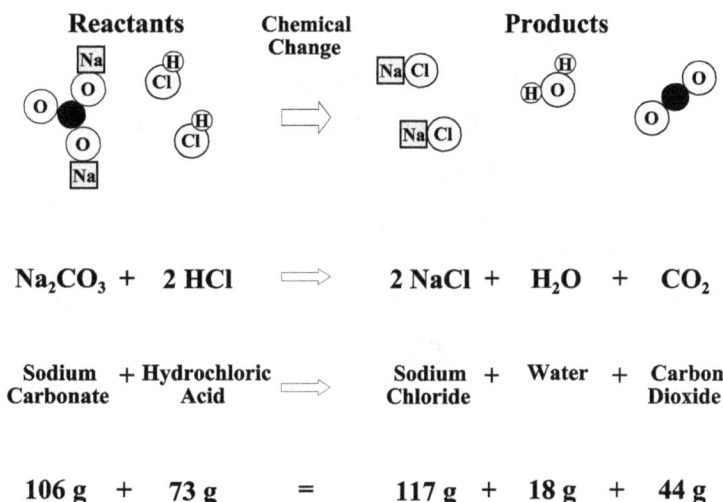

Reactants   **Chemical Change**   **Products**

$$Na_2CO_3 \ + \ 2\,HCl \quad \Longrightarrow \quad 2\,NaCl \ + \ H_2O \ + \ CO_2$$

Sodium   + Hydrochloric   Sodium   + Water + Carbon
Carbonate   Acid   Chloride   Dioxide

106 g   +   73 g   =   117 g   + 18 g + 44 g

Explaining Chemical Processes

The subject of chemical reactions is frequently introduced as a set of four classes: synthesis, decomposition, single displacement and double displacement.

The reactions in this unit have been chosen to represent these four classes.

On the other hand, this classification system has only limited descriptive value, and very little predictive or explanatory value. These exercises will therefore not emphasize this classification system as a learning outcome.

At room temperature, an air molecule's average speed is about 450 m/s. At 100°C, the speed is closer to 530 m/s. Water molecules, being lighter, have even greater speeds: about 600 m/s and 700 m/s respectively.

Learning Expectations CP1.01: recognize the relationships among chemical formulae, composition, and names; **1.02**: explain, using the law of conservation of mass and atomic theory, the rationale for balancing equations; **1.03**: describe, using their observations, the reactions and products of a variety of chemical reactions, including synthesis, decomposition, and displacement reactions; **2.08**: represent simple chemical reactions using molecular models, word equations, and balanced chemical equations; **2.09**: compare theoretical and empirical values and account for discrepancies when investigating conservation of mass.

Pedagogical Issues

The principal objective in these labs is the cultivation of the student's ability to represent chemical change in four parallel ways: Dalton diagrams, balanced chemical equations, word equations, and mass equations. Accordingly, the student exercises require that each and every reaction is described in all four ways. Students are also required to calculate the masses of reactants and products in every case, and to measure them in most labs. In this way, the students are confronted daily with evidence in support of Dalton's Theory of Chemical Processes. Note that the evidence they collect is not likely to be exactly the ideal values. Students, like practicing scientists, will have to deal with the question "what degree of agreement with prediction will I find to be convincing"? They are also gaining experiences that will strongly support the learning of the mole concept in Grade 11.

Science Issues

How can science teachers introduce some sense of the dynamics of chemical reactions, without the use of the advanced concepts such as rate, bond energy, and equilibrium? Students can comprehend the interplay between the *strength of the attractions* holding the atoms together in molecules, on the one hand, and the *thermal energy* tearing the molecules apart on the other.

Both $(NH_4)_2CO_3$ and $CuSO_4 \cdot 5 H_2O$ are torn into smaller pieces as the speeds of the molecules begin to reach 500, 600 and even 700 m/s. Ammonium carbonate is unusual in that all of the pieces are small enough to dissipate as gases.

Decomposition and Synthesis

"A rose by any other name..."

Let's see what happens when you pronounce the word "molecule" with a long "o" as in

"Hole - a - cule".

One "mole-cule" of ammonium carbonate has a mass of 96 g and decomposes into two "mole-cules" (34 g) of ammonia, one "mole-cule" (18 g) of water and one "mole-cule" (44 g) of carbon dioxide.

Next year, in Grade 11 chemistry,

just drop the "cule."

The Learning Activity 1.1: $(NH_4)_2CO_3$

Before the experiment, students will

Predict: which particles, if any, will break free of the ammonium carbonate molecule.

Explain: their prediction using words and pictures.

After the experiment, students

Observe: the appearance of the products, including the results of chemical tests.

Explain: what happened, using pictures and sentences.

Equipment, Preparation and Resources for each group

- test tube, Bunsen burner, test tube clamp, retort rod
- 1 g ammonium carbonate.
- access to a balance capable of 0.01 g precision
- dried cobalt chloride paper;   acid / base test paper

The Learning Activity 1.2: $CuSO_4 \cdot 5\,H_2O$

Before the experiment, students will

Predict: the mass of product.

Explain: the predictions using words and pictures.

After the experiment, students

Observe: products, results of chemical tests.

Explain: what happened, using pictures and sentences.

Equipment, Preparation and Resources for each group:

- test tube, Bunsen burner, test tube clamp, retort rod
- 2.5 g copper sulphate pentahydrate
- access to a balance capable of 0.01 g precision.
- dried cobalt chloride paper; water.

Be sure to keep all copper sulphate for future use.

Categories:	Assessment and Evaluation
Knowledge and Understanding:	Quality of representation of reactions, interpretation.
Thinking and Inquiry:	Quality of prediction and explanation
Communication:	Quality of argument, explanation
Applications / Connections:	

Explaining Chemical Processes

Lab 1.3: Using Iron to find % Oxygen in Air

Question: How can a teacher turn a routine synthesis reaction into an inquiry?

Answer: Use the synthesis reaction to answer another pressing question.

How strongly does oxygen attract iron's electrons?

Try this one:

Fluff up a small sample of steel wool so that you can see through the strands. Place it on a non-combustible surface. Apply a flame, and the steel burns brightly with considerable evolution of heat, producing Fe_3O_4.

The relative radius of oxygen and iron are provided in the diagram below.

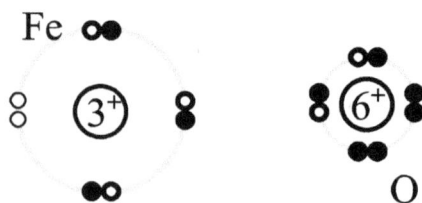

Fe

Oxygen will clearly grab iron's electrons away, thus combining with the iron to form oxide.

Learning Expectations CP1.01: recognize the relationships among chemical formulae, composition, and names; **1.02**: explain, using the law of conservation of mass and atomic theory, the rationale for balancing equations; **1.03**: describe, using their observations, the reactions and products of a variety of chemical reactions, including synthesis, decomposition, and displacement reactions; **2.08**: represent simple chemical reactions using molecular models, word equations, and balanced chemical equations; **2.09**: compare theoretical and empirical values and account for discrepancies when investigating conservation of mass.

Pedagogical Issue: Student use of evidence and argument

There are several ways to synthesis oxides of iron. In general, they do not provide easily measured reactants and products. Important evidence on the formula of the product is impossible to collect in the lab; the products are complex and impure; some iron fails to react, some leaves the scene as smoke, etc. Just telling kids what the results ought to be is a very distant second best.

But... by using the synthesis of Fe_2O_3 to trap the O_2 in air, we can measure the % contribution of oxygen to the atmosphere. This is a meaningful problem to students. The most important evidence is relatively simple to collect, and the technique demonstrates a widely used method for determining an unknown quantity. Furthermore, some students will, on the basis of their evidence, develop defensible positions that are at odds with the rest of the class.

Science Issues Fe_2O_3 initially forms only in the presence of water, and probably includes water in its crystal. When placed in contact with water, Fe_2O_3 is almost insoluble. If Fe^{3+} ion comes into contact with water, it quickly sequesters hydroxide into an insoluble solid, and releases H^+ ion. In other words, iron, like many transition metal ions, forms acidic solutions by the process of hydrolysis. You may not get the clear metal oxide forming base evidence that you expect.

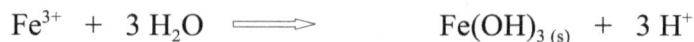

$$Fe^{3+} + 3\ H_2O \implies Fe(OH)_{3\ (s)} + 3\ H^+$$

An important pedagogical / scientific issue is the means by which students represent air. The use of the particle theory, and particle diagrams, can enhance student understanding. As the oxygen particles are removed from the gas state, only the other gases remain (mostly nitrogen). The water is forced up to occupy the space that was formerly filled by the oxygen molecules.

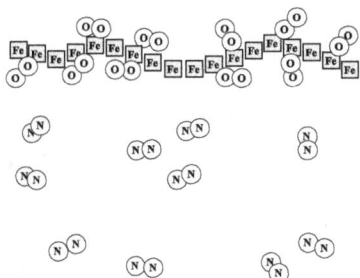

Best measurement of the volume of oxygen in air is obtained if the students adjust the depth of the graduated cylinder so that the water level inside the cylinder is the same as the water level outside the cylinder. This will reduce the effect of pressure changes inside the cylinder.

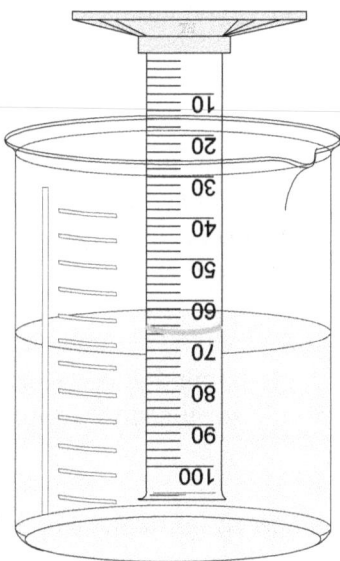

The Learning Activity

Before the experiment, students will

Predict: the percentage of the earth's atmosphere that is oxygen, O_2..

Explain: the prediction in diagrams and sentences.

After the experiment, students

Observe: all changes, including appearances of the iron, as well as the volume of gas.

Explain: their observations in diagrams and sentences.

Equipment, Preparation and Resources

- clean dry steel wool
- alcohol and 1.0 M HCl to clean the steel wool
- 100 ml graduated cylinder
- 250 ml beaker

Disposal:

Wet steel wool will continue to oxidize in the presence of air, generating considerable heat. To avoid the possibility of fire or other reactions, either submerge the steel wool in water until it is oxidized, or dry it out before disposal. *Steel wool wrapped in a slightly damp paper towel can get quite hot, especially if the paper towel contains an electrolyte like salt or HCl.*

Categories:	Assessment and Evaluation
Knowledge and Understanding:	Quality of student representations of the reactions
Thinking and Inquiry:	Quality of the student predictions and explanations
Communication:	Quality of student explanation
Applications / Connections:	This lab is an application of a chemical reaction

Explaining Chemical Processes

Question: why is it that when you burn something at home, the mass of the ash appears to be less than that of the fuel. But when we burn something here at school, the mass appears to be greater?

Is there something about schools that makes ash heavier?

Students have a clear notion of "domains" of knowledge: real knowledge is good in the real world, while school science knowledge is only good in schools.

Students can easily believe that the funny metal stuff you get in schools produces ash that is heavier, but that doesn't happen in the real world.

Whenever we use the term "real world" in our discourse, we may be seriously undermining our role as teachers.

Lab 1.4: Synthesis of Magnesium Oxide

Learning Expectations CP1.01: recognize the relationships among chemical formulae, composition, and names; **1.02**: explain, using the law of conservation of mass and atomic theory, the rationale for balancing equations; **1.03**: describe, using their observations, the reactions and products of a variety of chemical reactions, including synthesis, decomposition, and displacement reactions; **2.08**: represent simple chemical reactions using molecular models, word equations, and balanced chemical equations; **2.09**: compare theoretical and empirical values and account for discrepancies when investigating conservation of mass.

Pedagogical Issues

It is widely believed by teachers that, given two competing accounts of a phenomenon, students will hold to the most vividly experienced account. One of the most vivid experiences of a young person is that burning reduces substances to a small quantity of ash. Even after being provided with evidence, students find it difficult to use the idea that when a metal reacts with oxygen, the metal oxide has greater mass than the original metal. In addition to providing a good example of a synthesis reaction, this lab provides an opportunity for students to measure the change in mass as oxygen atoms are incorporated into magnesium metal.

Science Issues

This lab is probably best conducted as a demonstration. Two or three samples of Mg can be oxidized simultaneously in a fume hood or on the demonstration lab bench. The reaction is usually quantitative, but a few precautions will help provide good results. Take time to ensure that the Mg ribbon is:
- carefully cleaned before weighing
- coiled loosely near the bottom of the crucible that is covered with a lid, with a small crack to provide for the slow admission of air to the mixture.

Heat must be applied carefully - enough heat to ignite the Mg in the crucible, and to sustain the reaction as O_2 becomes more scarce. As the reaction slows and even nearly stops, the lid may be removed with tongs to allow complete reaction. Don't let any smoke escape!

The crucible must be carefully cooled after the reaction is complete.

Decomposition and Synthesis

The mass of the products is not likely to be exactly the ideal value, that is, 0.40 g.

It should, on the other hand, be close enough to be convincing.

What does that say about the exactness of scientific reasoning?

Is the aim of this lesson to measure the exact mass of MgO produced or to have kids reason for a short time much as a scientist does?

Is the reasoning of experienced scientists in the field exactly the same sort of thing as the reasoning of novices as they become familiar with science?

If not, then what allowances must we make in science classes?

Categories:
Knowledge and Understanding:
Thinking and Inquiry:
Communication:
Applications / Connections:

The Learning Activity

Before the experiment, students will

Predict: the mass of magnesium oxide (MgO) that can be formed from 0.24 g of Mg metal.

Explain: the prediction in diagrams and sentences.

Students will conduct the following experimental procedure:
1. Weigh a clean dry crucible and lid.
2. Cut a piece of Mg ribbon with a mass of 0.24 g. Coil it into a spiral, and put it into a crucible.
3. Measure the mass of the crucible + lid + magnesium.
4. Put a lid on the crucible at a slight angle. Gently heat the crucible for 5 min, then heat to red hot for about 10 min.
5. Carefully open the lid about 1 mm with tongs. If smoke escapes, close lid . Heat another 5 min, gradually removing lid.
6. Allow the crucible to cool for 15 min.
7. Weigh the crucible + lid + product MgO.
8. Examine the product.

After the experiment, students

Observe: all changes, including appearances of the magnesium and the product.

Explain: their observations in diagrams and sentences.

Equipment, Preparation and Resources

For each determination, you need
- 1 crucible + lid
- Bunsen burner, iron ring, crucible tongs, clay triangle, lighter
- 0.24 g of Mg ribbon
- Balance capable of reading 0.01 g

Assessment and Evaluation

Quality of student representations of knowledge
Quality of questions, predictions, and applications
Quality of writing, diagrams, etc.

Explaining Chemical Processes

Lab 2.1: Reactions of Non-metals and Oxygen

The Dalton diagrams and chemical equations learned previously can be applied here to support a very simple generalization:

"Non-metal oxides form acidic solutions in water."

The series of transitions:

- from non-metal to oxide
- from oxide to solution
- from solution to compound
- from compound to ions

is quite complex, and certainly not obvious to a student. Each step is important in the overall picture. Does the student have enough familiarity with each step to grasp the nature of each transition?

Learning Expectations CP 1.03: describe, using their observations, the reactants and products of a variety of chemical reactions, including synthesis, decomposition, and displacement reactions; **1.05**: explain the interrelationships among metals and non-metals, acidic and basic oxides, and acids, bases, and salts; **2.11**: conduct experiments on the combustion of metals and non-metals and react the oxides formed with water to produce acidic or basic solutions.

Pedagogical Issues

Students are unlikely to have "everyday ideas" about something so far from everyday experience as this lab is. The ideas that we want our students to learn here are more abstract than everyday thinking would be. We propose to frame the learning outcomes in two distinct ways:

1. In this lab, the learning is limited to the student's ability to recite, on the basis of limited experience, that non-metal oxides form acidic solutions. This is generally accepted as an achievable goal. Having limited predictive and almost no explanatory power, however, this goal may not be a good example of "science knowledge".

2. In the next activity, the objective is to explain *why* non-metal oxides form acidic solutions.

3. In the third activity, the student extends the representation to a large class of acids.

Science Issues

Students are already familiar with the first step in this reaction, a simple synthesis reaction. The second step is the formation of ions.

The student is likely to have a very limited understanding of ions and electrolytes at this stage. To what degree do you wish to bring the issue up? It is possible to demonstrate the existence of ions by showing that the solution becomes conductive when it becomes acidic.

Metals, Non-metals, and pH

Why do non-metal oxides form acidic solutions?

Because the student is unlikely to have a suitable mental model, the student's first answer to this question must be a circular argument, i.e., a tautology: "The solution is acidic because it is a non-metal. Non-metal oxides form acid solutions".

The student is unlikely to be able to proceed further. But this provides us with a problem in science education. Tautologies are not science, even when they refer to science concepts. Your most perceptive students are instinctively unsatisfied with a tautological definition. Your weaker students, on the other hand, will recite the circular argument until the exam, and then forget it entirely. In the discipline of science, we must explain why an acid is an acid. Furthermore, we must explain why a non metal oxide invariably forms an acid solution.

The Learning Activity

Before the experiment, students will

Predict: Will the solution be acidic or basic?

Explain: the prediction using words and diagrams.

After the experiment, students

Observe: the colour change in the indicator.

Explain: why the solution became acidic.

Equipment, Preparation and Resources

For each lab group, you will need:

- 2 prepared gas bottles containing oxygen for each lab group. Each bottle must have 1 cm indicator solution at the bottom.
- 1 deflagrating spoon for each group
- small pieces of sulfur, about the size of a grain of rice.
- lumps of charcoal
- Bunsen burner

The indicator solution can turn colour prematurely if there is any residue in the gas bottles. This can be minimized by cleaning the gas bottles before use.

Mix the indicator solution before adding it to the bottles. Start with one litre of tap water, and add enough bromthymol blue to be easily visible. The colour is probably green, but may be slightly yellow or blue.

If you intend to test the student's acid solutions for conductivity, test the indicator solution before you use it to be certain that it does not conduct electricity before the gaseous oxides are added.

Categories:

Knowledge and Understanding:
Thinking and Inquiry:
Communication:
Applications / Connections:

Assessment and Evaluation

Simple claim: non-metal oxides form acidic solutions

Clarity of representations
Application of previous representations

To understand the tug-of war for electrons, we use the core-valence-radius diagrams. The completed diagrams looks like this:

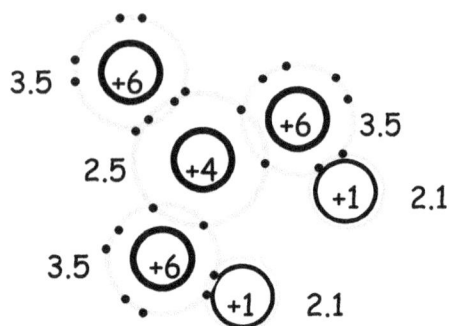

In the tug-of-war for electrons, the "big winner" is oxygen: electrons will be strongly attracted to its +6 core charge. The "big loser" is hydrogen; even though it has the smallest radius, it has by far the least core charge. Carbon falls in the middle.

Comparing electronegativities, we reach the same conclusion. Oxygen is the big winner, hydrogen is the big loser. Carbon falls in the middle.

If hydrogen loses its electrons, it becomes H^+.

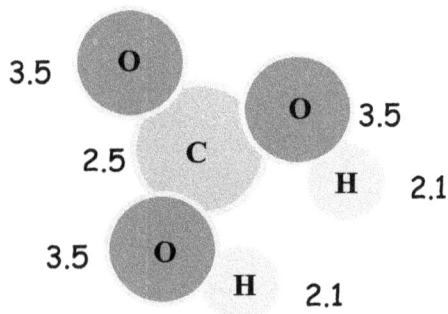

Activity 2.2: What Makes Acids Acidic?

Learning Expectations CP 1.03: describe, using their observations, the reactants and products of a variety of chemical reactions, including synthesis, decomposition, and displacement reactions; **1.05**: explain the interrelationships among metals and non-metals, acidic and basic oxides, and acids, bases, and salts; **2.11**: conduct experiments on the combustion of metals and non-metals and react the oxides formed with water to produce acidic or basic solutions.

Pedagogical Issues
We will be making use of two features of the student's thinking here. Kids (and adults) rapidly size up "winners" and "losers", given an appropriate set of cues. This ability is not limited to the human race, and is likely a very important cognitive strategy for survival.

If we chemists supply cues that are both scientifically sound, and psychologically significant, our students can make scientifically valuable judgements using this strategy.

The core-valence-radius table provides a set of cues that are consistent with the nature of scientific knowledge. Within that framework, the concept of electronegativity can have meaning for an adolescent.

Science Issues
It is not necessary at this point that the students have the ability to construct representations of covalent bonding. It is valuable that, presented with such diagrams, students can decide which way the electrons will tend to move.

Hydrogen's electronegativity is very nearly that of the metalloids, that is 2.0 ±0.2. All of the non metals, then, have ε greater than hydrogens. All of the metals have ε less than that of hydrogen. This simple fact is the key difference between the behaviour of the oxides of metals and non-metals in aqueous solution. Of course, electronegativity is itself a complex property, arising from a number of other systematic influences.

Metals, Non-metals, and pH

For chemistry teachers, this pattern can be extended to find the relative acidity of many compounds. Why, for instance, is sulfurous acid H_2SO_3, weaker than sulfuric acid H_2SO_4, ?

Let's compare a simple product of the electronegativities of the O and S atoms in each molecule

$$(3.5)^3 \times 2.5 = 107$$
for H_2SO_3

$$(3.5)^4 \times 2.5 = 375$$
for H_2SO_4

This is a crude, but understandable, entry to a discussion of acid strength.

We are using the students' schematic reasoning, and a few accessible numbers, to arrive at a rough description of acid strength.

Traditional approaches, using formal mathematical definitions, are not within

The Learning Activity

Students compare the acidity of acids with the following instructions:

1. Write the core charge in the core of each atom.

2. Determine the electronegativity ε of each atom in the acid molecule, using the core - valence - radius table. Write the ε near each atom.

3. In the tug-of-war for electrons, who is "the big winner"? Which atom will attract electrons most strongly?

4. In the tug-of-war for electrons, who is "the big loser"? Which atom will lose its electrons to the others?

5. If "the big loser" in each case loses its electrons, what does it become?

Equipment, Preparation and Resources
- exercises in the student manual
- pens, pencils etc.
- access to the core -valence-radius version of the periodic table

Categories:
Knowledge and Understanding:
Thinking and Inquiry:
Communication:
Applications / Connections:

Assessment and Evaluation
Understands acid strength and relationship to core-valence-radius
Quality of students written explanations

Recognizes applications of understanding of acid strength

Explaining Chemical Processes

Of all the elements in the periodic table, fluorine and oxygen have the greatest ability to attract and hold electrons from other atoms. This ability is called electronegativity, ε.

Oxygen, ε = 3.5, is the second most electronegative element in the universe, and will attract electrons from all other elements except fluorine.

The systems we are discussing have only three elements: oxygen, a second non-metal, and hydrogen.

Oxygen and the second non-metal will always take the electron from hydrogen.

Therefore, non-metal oxides invariably form acidic solutions.

Activity 2.3: Reactions of Non-metal Oxides to form Acids

Learning Expectations CP1.01: recognize the relationships among chemical formulae, composition, and names; **1.02**: explain, using the law of conservation of mass and atomic theory, the rationale for balancing equations; **1.05**: explain the interrelationships among metals and non-metals, acidic and basic oxides, and acids, bases, and salts; **1.08**: name and write the formulae of common ionic and molecular compounds **2.11**: conduct experiments on the combustion of metals and non-metals and react the oxides formed with water to produce acidic or basic solutions.

Pedagogical Issues

This is the first extensive student exercise in using formal balanced equations. It also introduces the most complex nomenclature issues to date. We introduce nomenclature with no explanation. The students will be required to write each name at least once, within the context of a balanced chemical equation. The name itself will be only a few centimetres from the chemical formula.

The primary learning outcome is the recognition that non-metal oxides form acidic solutions. The nomenclature is a secondary outcome, but an important one. In this exercise, the student is provided enough support to begin to recognize a few naming conventions. We make no attempt here to formalize the naming conventions, although you may wish to do so in your own class.

Repetition of the formal balanced chemical equation and the names reinforces three *conceptual* steps:
  i)   Non-metals react with oxygen to form non-metal oxides;
  ii) Non-metal oxides react with water to form acids;
  iii) Acids dissolve in water, dissociating into hydrogen ions and polyatomic ions.
In a real situation, the reactions are not so simple, and other products can form at each "step".

Science Issues

All of the non-metal oxides are named according to the Greek prefix system. The acids are named according to the accepted "ic / ous" system. None of the "hypo" or "per" acids are described.

Demonstration:
Generate a small amount of N_2O_4 in a clean, dry 250 mL Erlenmeyer flask:

Put a penny into the flask.
Add 5 - 10 drops of concentrated nitric acid.

The flask will be filled with a cloud of dark brown nitric oxide N_2O_4.

N.B. Nitric oxide is noxious. Nitric acid is dangerous. Quantities must be carefully limited. The demonstration must be carried out in a fume hood. Students must not conduct this experiment.

"Pour" the dense, brown cloud of nitric oxide N_2O_4 gas into a large beaker with some water and bromthymol blue indicator. The gas rapidly dissolves in water, turning the bromthymol blue to yellow.

NB. The dissociation of acids into ions is being introduced *before* formal instruction on the formation of ions. You may need to elaborate this point to support student understanding.

The Learning Activity
Students work in pairs or small groups to:
1. Balance all of the equations in the list in the student exercises.
2. Fill in the blanks with the names of the substances in each equation.

Note that these reactions are set up in groups of three. The three equations are not *temporally* sequential, but they are *conceptually* sequential. It is easy to imagine, for example, dissolving without dissociation, but it is much more difficult to imagine dissociation without dissolving.

Each equation leads naturally to the next. The products of the first reaction are the reactants of the second, and so on. This is also true for the names. Your students will quickly figure out how to exploit this system to write the names in the blanks.

This activity may be extended by requiring the use of Dalton pictures. The concept of acidity may be further reinforced by having students conduct analyses of the electronegativities, as in the previous exercise.

Equipment, Preparation and Resources
Student exercises, pens, pencils, etc.

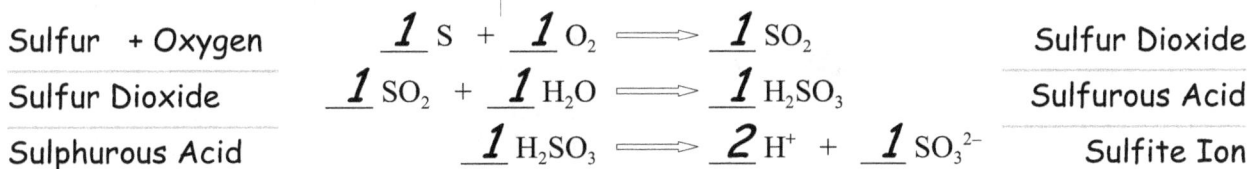

Sulfur + Oxygen	1 S + 1 O_2 $\longrightarrow$ 1 SO_2	Sulfur Dioxide
Sulfur Dioxide	1 SO_2 + 1 H_2O $\longrightarrow$ 1 H_2SO_3	Sulfurous Acid
Sulphurous Acid	1 H_2SO_3 $\longrightarrow$ 2 H^+ + 1 SO_3^{2-}	Sulfite Ion

Categories:	Assessment and Evaluation
Knowledge and Understanding:	Recognizes and writes correct names and formulae
Thinking and Inquiry:	Balances equations; predicts acid behaviour
Communication:	Uses correct names and formulas
Applications / Connections:	Extends knowledge to new situations

Explaining Chemical Processes

The core-valence-radius diagram of MgO and H_2O looks like this:

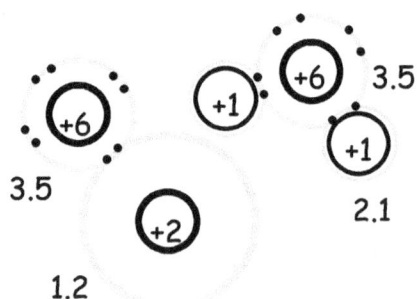

The "big winner" of electrons is, as always, oxygen. Electrons will clearly be strongly attracted to its large core charge and relatively small radius. The "big loser" however, is now magnesium. With its large radius and small core charge, it simply cannot hold on to its own electrons in the presence of O and H.

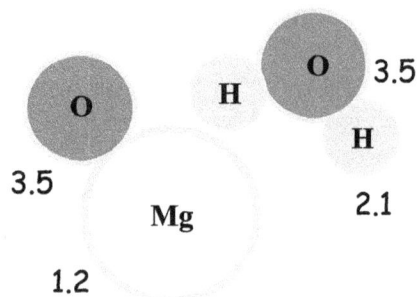

It is clear that Mg will lose its electrons and form the Mg^{2+} ion. Hydrogen "falls in the middle" in terms of ε, and forms a covalent bond in the OH^- ion.

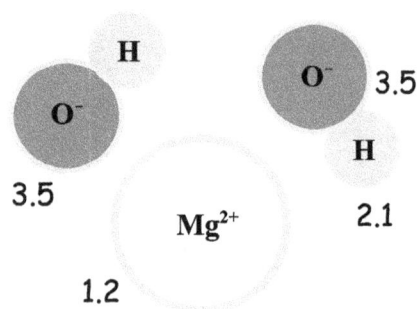

Activity 2.4: Metal Oxides and Bases

Learning Expectations CP 1.01: recognize the relationships among chemical formulae, composition, and names; **1.02:** explain, using the law of conservation of mass and atomic theory, the rationale for balancing equations; **1.05:** explain the interrelationships among metals and non-metals, acidic and basic oxides, and acids, bases, and salts; **1.08:** name and write the formulae of common ionic and molecular compounds; **2.11:** conduct experiments on the combustion of metals and non-metals and react the oxides formed with water to produce acidic or basic solutions.

Pedagogical Issues and Science Issues

This set of concepts is very similar, though not identical to, the non-metal oxide topic covered in the past three lessons. Once again, we invoke the students' notion of "strength" to predict "winners and losers". This pedagogical strategy provides deep and immediate meaning in the cognitive world of learners.

It can only be honest to science, however, if we teachers provide the students with an appropriate representation of atoms. The representation must have two properties. First, it must be coherent with the larger scientific community's understanding of the atom. Second, it must be amenable to being utilized by students within their natural schematic thinking.

In this case, we use the core-valence-radius model of the atom, and its logical corollary, electronegativity. The "strength" of electronegativity can be used by students to predict "winners" and "losers" of electrons.

Electronegativity itself can be understood in terms of core charge and atomic radius. Thus the whole construct is accessible to ordinary students.

Metals, Non-metals, and pH

Demonstration:

Put 1 cm of water plus indicator (bromthymol blue or universal indicator) into a clean gas bottle.

Take a small piece (4 cm) of Mg metal in tongs, ignite the Mg in a Bunsen burner flame, and hold the burning Mg over the mouth of the gas bottle.

As the MgO forms, and falls into the indicator solution, the indicator turns blue, indicating the presence of base.

Other metal oxides will also form bases. Try CaO and ZnO.

NB. Many transition metal oxides do not form strong bases!! Instead, they "split water" into H^+ and OH^-, and then form insoluble hydroxides, leaving the solution acidic.

The Learning Activity
Students work in pairs or small groups to:
1. Balance all of the equations in the list in the student exercises.
2. Fill in the blanks with the names of the substances in each equation.

Once again, the equations are grouped into sets of three:
   i)   formation of the oxide;
   ii)   reaction with H_2O to form the hydroxide;
   iii) dissociation of the hydroxide.

In every case, the product of the leading equation is the reactant for the equation that follows. This should have the students writing a large number of chemical names and formulas in a short period of time.

Equipment, Preparation and Resources
For the demonstration:
- Magnesium ribbon, Bunsen burner, tongs, gas bottle, water, indicator solution

For the student exercise:
- Pens, pencils, etc, and the student exercises

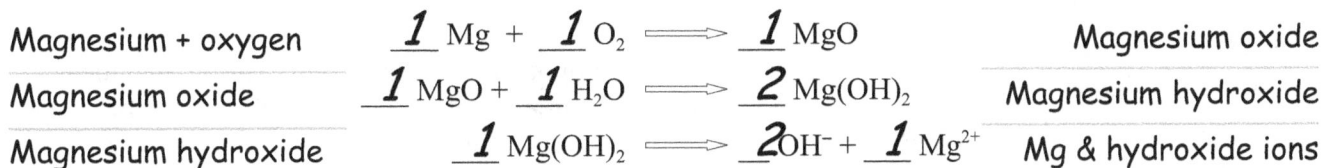

Magnesium + oxygen	$\underline{1}\ Mg + \underline{1}\ O_2 \Longrightarrow \underline{1}\ MgO$	Magnesium oxide
Magnesium oxide	$\underline{1}\ MgO + \underline{1}\ H_2O \Longrightarrow \underline{2}\ Mg(OH)_2$	Magnesium hydroxide
Magnesium hydroxide	$\underline{1}\ Mg(OH)_2 \Longrightarrow \underline{2}OH^- + \underline{1}\ Mg^{2+}$	Mg & hydroxide ions

Categories:	Assessment and Evaluation
Knowledge and Understanding:	Correctly predicts the name and formula of the products
Thinking and Inquiry:	Explains the formula and name choices
Communication:	Spelling; writing balanced equations
Applications / Connections:	Draws appropriate Dalton diagrams

Explaining Chemical Processes

High school students (and adults) have a profound cognitive commitment to the idea that "double the effort doubles the results". This simple pattern is one of the schematic structures that is widely observed in all ad hoc human thinking.

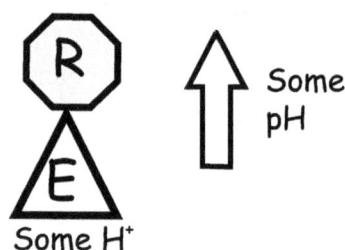

When talking about acid, kids are likely to think that pH should be directly proportional to the H⁺ concentration.

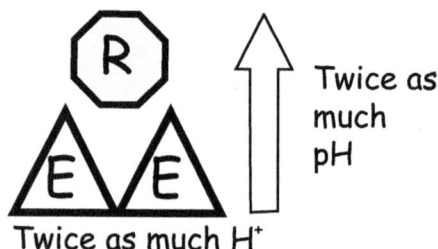

Of course, this relationship does not hold. Because this idea is so intuitively meaningful, kids will find it difficult to abandon.

Lab 2.5: Testing the Strength of Acids and Bases

Learning Expectations CP 1.07: describe how the pH scale is used to identify the acidity of solutions; **1.06**: describe qualitatively acid-base neutralization through observation of simple acid-base reactions; **2.10**: conduct experiments to identify the acidity and basicity of some common substances.

Pedagogical Issues and Science Issues

The formal definition of pH is:

$$pH = -\log_{10}[H_3O^+]$$

This definition is beyond the reach of students at this point for several reasons.

1. They do not have a working quantitative definition of concentration, let alone molar concentration.
2. The significance of the H_3O^+ ion has not been cultivated.
3. They do not have a working knowledge of logarithms, nor of the nature of a log function.

Because each of these ideas requires considerable effort to learn, this problem cannot be realistically resolved by attempting to teach logs, or molar concentration.

That leaves teachers in a difficult position. In the absence of the traditional supports, we must convey some sense of the pH scale. We must aim for a more direct experiential understanding of pH at this point. The most accessible ideas we have to work with are:

1. Our students likely have some sense of dilution and of relative concentration.
2. Our students can make pictorial and text representations of a "factor of 10" dilution.

The diagram on the opposite page provides a simple graphical representation of the increments of concentration in the pH scale. Each step represents one order of magnitude. There is one major difficulty with this picture: water molecules are under-represented. One litre of water contains 55.5 mol of water molecules. In a 1 M H⁺ solution, each H⁺ ion would be outnumbered by about 55 molecules of H_2O. In order to represent the water and acid particles proportionally, we would have to add 5500 molecules of water to each of the beakers.

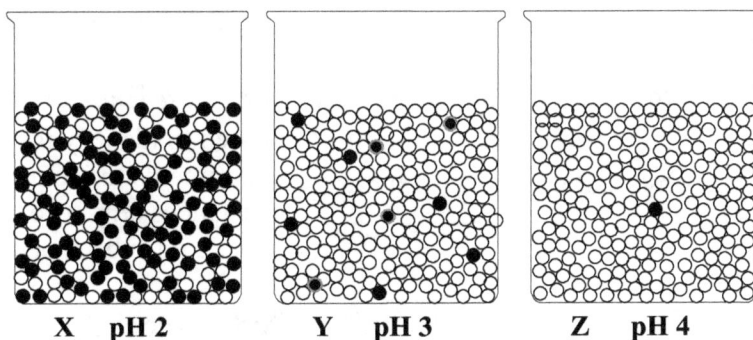

X   pH 2   Y   pH 3   Z   pH 4

The students will learn the technique as they proceed.

It will take approximately 10 mL of 0.010 M HCl to bring the solution to pH 3. This will correspond to about 100 - 200 drops!

There are a number of factors which compromise the results of this experiment:

Not all drops are the same size.

Not all experiments are diluted to the same extent. You have to add about 10 mL of water (containing HCl, of course) to the pH 3 sample, but only 2 drops of water to the pH 5 sample.

Nevertheless, the general idea that one pH point is equivalent to 10 times more acid is easily demonstrated

The Learning Activity

Students complete the following procedure:

1. Obtain 500 mL of distilled water. Add 5 mL of Universal Indicator. The solution should be neutral (pH 7.0, yellow-green).
2. Measure 100 mL of the neutral solution into 5 clean dry beakers and label the beakers pH 3, pH 4, pH 5, pH 6 and pH 7.
3. Obtain 100 mL of strong acid, pH 2.0 from your teacher.

Predict how many drops of the strong pH 2 acid you must add to each beaker to make the pH indicated. **Explain** your prediction.

4. One drop at a time, counting each drop, add the strong pH 2 acid to the neutral water to make pH 6. *Stir after adding each drop.*
5. Repeat the procedure to make solutions pH 5, pH 4 and pH 3.

Observe the number of drops and record on the diagram.
Explain your results.

Equipment, Preparation and Resources

For each lab group you will need:
- 500 mL distilled water
- 5 mL universal indicator
- 100 mL 0.010 M Hcl
- 6 250 mL beakers
- disposable dropper

Categories:

Knowledge and Understanding:	**Assessment and Evaluation**
Thinking and Inquiry:	Describes adjacent pH points as being 10× more acidic.
Communication:	Nature of students reasoning as they work through the data
Applications / Connections:	Clarity of writing and diagrams

Explaining Chemical Processes

Lab 2.6: Testing the pH of Common Substances

Try to recall a time when you learned to measure a new physical quantity, perhaps mass or volume.

Can you recall how many things you attempted to measure at that time? The more things you measured and compared, the more coherent your understanding became.

Unfortunately, when you or your friends stopped performing this new measurement, the learning began to fade, as new neural connections were no longer being reinforced.

Young people are very familiar with a large number of substances around their homes. This exercise provides an opportunity to measure familiar things with a new scientific scale.

Even though the pH paper strips are expensive, be as generous as you can reasonably afford to be. The opportunity for your students to reinforce this learning at a later date will not come soon.

Learning Expectations CP 1.07: describe how the pH scale is used to identify the acidity of solutions; **2.10:** conduct experiments to identify the acidity and basicity of some common substances (e.g., use acid-base indicators to classify common household substances according to the pH scale).

Pedagogical Issues

There are two distinct representational issues here. The first is the identification of specific substances with particular pH values. It is unlikely that students will recall the pH of every substance that they test. It is likely that they will recall the pH of at least one common household substance. This kind of knowledge is not systematic, and is very difficult to accurately recall later.

The second representational issue is the relationship between adjacent pH values. Note that students will do this twice for each substance tested. This kind of knowledge is highly systematic, and is much more likely to be remembered by students. It is, therefore, by far the most important part of the lab exercise for future learning.

Science Issues

The scientific meaning of "strong acid" is "an acid which dissociates completely". A "weak acid" is "an acid which dissociates only partially". We cannot easily make such a distinction at this time and yet, we can hardly escape using the term "strong acid" and "strong base" incorrectly in this lab, because the student understanding of "strong" and "weak" are simply impossible to avoid. We must bow to the limits of student cognition at this time, and put the science on the back burner for a few years.

The only accommodation we can make at this time is to relate the term "strong acid" to the relative concentration of H^+ ion. This is yet another reason to require students to express the relative number of H^+ and OH^- ions in every case.

Metals, Non-metals, and pH

Some practical hints:

Every student will need a colour pH scale that matches that of the pH paper strips. The best solutions:

Keep all of the pH scales. Don't throw them away when the packages are empty.

Scan one colour pH scale and pick it up as a graphic file. Make multiple copies of the graphic file onto one page, and print the page on a colour printer.

The Learning Activity

Students will complete the following procedure:

1. Make a list of 20 substances found in your house. **Predict** the pH of each substance.

2. **Explain** your prediction, e.g. *"I think that milk has 1/10 as much H^+ as vinegar, so it will have a pH of ____ ."*

3. **Observe** the pH by testing with universal indicator or pH paper.

4. **Explain** your measurements, e.g.. *"Milk has a pH of ____ , vinegar's pH is ____ . Therefore milk has ____ × more H^+ than vinegar.*

For example, in actual measurements, a student might find:

Milk is   pH 7.
Vinegar is   pH 3.
Therefore, vinegar has 10 000 × more H^+ than milk.

Equipment, Preparation and Resources

Each student will need enough pH paper to test 20 samples. Ten cm of paper would be *just* enough.

Categories: Assessment and Evaluation

Categories:	Assessment and Evaluation
Knowledge and Understanding:	Relates differences in pH to differences in concentration
Thinking and Inquiry:	Chooses an appropriately broad set of materials to test
Communication:	Clarity of writing and expression
Applications / Connections:	The overall exercise is an application of pH

This is a most dramatic reaction: Two dangerous, exotic, toxic substances combine to make common, benign household salt!

The most accessible test for salt is, of course, to taste it. But what about safety?

Purple cabbage juice is a wonderful indicator which undergoes a colour change near pH 7. Purple cabbage is:

purple   @ pH 6
navy blue   @ pH 7
teal blue   @ pH 8

Even if the acid and base are not perfectly neutralized, students will tend to fall between pH 5 at the lower limit (about as acidic as apple juice) and pH 9 at the upper limit (about as basic as baking soda). Most students will be much closer to pH 7 than this.

Furthermore, purple cabbage juice is non-toxic.

Bromthymol blue, phenolphthalein, and universal indicator are not safe, and definitely should not be ingested.

Lab 3.1: Acids, Bases and Neutralization

Learning Expectations : *CP 1.06*: describe qualitatively acid-base neutralization through observation of simple acid-base reactions.

Pedagogical Issues At this point in our understanding of human cognition, it is possible to claim that human intelligence is profoundly dependent upon our emotional system.

Our emotional-attentive system is what triggers our sense of significance and meaning, focuses our attention, and primes the neurons for permanent change (learning). How could we possess intelligence without that?

This lab has very strong emotional overtones, and will likely be one of the most memorable events in this class. It is very important that we direct students in a way that provides them with a safe environment in which to experience this emotionally laden event.

Science Issues When a solution is near neutral, pH 7, the addition of even tiny quantities of NaOH or HCl can push the pH much farther than the student intends.

For example, a single drop of 0.10 M NaOH will contain approximately 5×10^{-6} moles of NaOH. If this one drop is added to a small volume of weak acid, e.g. 50 mL of pH 6 acid, the concentration of OH^- would be nearly 1.0×10^{-4} Mols per litre, or approximately pH 10.

It is very difficult to reach exact neutrality. Two strategies can help avoid wide swings in pH, and end up with .

First, both the NaOH and the HCl solutions must be quite dilute, about 0.10 mols per litre. This will result in a final NaCl concentration of about 0.050 M, or about 2.8 grams of NaCl per litre. This solution is just easily concentrated enough to taste the salt.

Reactions of Acids and Bases

Purple Cabbage Extract (PCE)

Chop a head of purple cabbage up into small pieces, like cabbage salad, and put it into a large clean beaker.

Add enough distilled water to just barely cover the cabbage.

Heat the water to boiling for 10 minutes, or until the water is deep purple. Let cool.

Pour off the coloured extract.

This solution is typically just a little acidic, about pH 6.

Prepare one sample of the PCE with pH 7 buffer, to demonstrate the colour of the PCE at neutral pH 7.

Second, the concentration of salt can be increased by evaporating the water to leave a solid residue. Rather than attempt this with a hot plate or open flame, we suggest that you place the beakers in an oven at $\approx 70\,^\circ C$ overnight. A slight excess of HCl is preferable to an excess of NaOH, as the NaOH will remain, but any excess HCl will evaporate harmlessly. The following day, the residue will be much more easily identified as common table salt.

The Learning Activity   Provide each lab group with:

>   25.0 mL of 0.10 M HCl
>   25.0 mL of 0.10 M NaOH
>   Purple cabbage extract indicator (PCE)

1.   Add 5 mL of PCE to each solution. Note the colours.

2.   Pour 20.0 mL of HCl into a 100 mL beaker. **Predict** the volume of NaOH needed to turn the solution neutral (navy blue). **Explain** your prediction.

3.   Use a dropper to add NaOH to the HCl solution in the beaker. Add small quantities at a time, and swirl. If you add too much NaOH, add a drop of HCl solution.

4.   When you have reached pH 7 (navy blue), taste a drop or two of the solution.

5.   Write **Observations** and **Explain** your results.

Equipment, Preparation and Resources   For each group:

25 mL 0.10 M HCl
25 mL 0.10 M NaOH
purple cabbage
2 test tubes, 1 100 mL beaker, 2 disposable droppers

Categories:

Knowledge and Understanding:
Thinking and Inquiry:
Communication:
Applications / Connections:

Assessment and Evaluation

quality of explanation of neutralization

quality of equations, diagrams, etc

Explaining Chemical Processes

Here's a dramatic titration that illustrates the reaction of H^+ and OH^- to form water.

Prepare:
dilute (0.010 M) $Ba(OH)_2$
dilute (0.010 M) H_2SO_4

Add universal indicator to both solutions. Put 100 mL of $Ba(OH)_2$ into a beaker, with a conductivity tester. The light will shine brightly.

Add the H_2SO_4 gradually.

As the reaction approaches the end point, the calcium and sulfate ions form insoluble calcium sulfate. The solution becomes neutral, and the light goes out as all of the ionic species disappear from the solution.

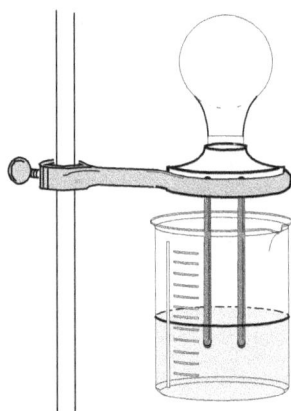

Lab 3.2: More Neutralization Reactions

Learning Expectations : *CP* **1.06:** describe qualitatively acid-base neutralization through observation of simple acid-base reactions.

Pedagogical Issues In the previous lab, we might have stated the generalization:

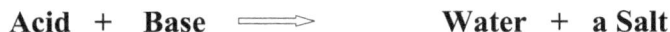

Acid + Base $\implies$ Water + a Salt

Chemists find this statement acceptable. For most students, however, the notion of "a salt" is problematic. Common table salt is the only salt with which most students are familiar. Some students may have difficulty expanding the meaning of the word "salt" to include any soluble ionic substance.

This lab provides two additional examples of the formation of salts. In both cases, the salts are insoluble, and quite visible.

Science Issues Both of these neutralizations are, of course, double displacement reactions. The products of the first reaction are water and calcium sulfate. The products of the second reaction are water and insoluble calcium carbonate.

Both of these double displacement reactions produce insoluble salts. As the ions are removed from the solution, the solution loses its electrical conductivity. It is possible to show that the electrical conductivity drops to a value near zero, as the neutralization point is approached. Conductivity returns as excess acid or base is added to move away from neutrality. The pH indicator should provide visible evidence that acid base neutrality has been achieved.

Reactions of Acids and Bases

Science and Pedagogy

The formation of calcium carbonate as Ca(OH)$_2$ limewater solution reacts with carbon dioxide, is, of course, the "test for carbon dioxide."

This test has been used in both grades nine and ten, without explaining the chemistry. Now the students can describe this simple chemical test in terms of a chemical reaction.

Of course, all chemical tests involve chemical change.

In fact, the only way to test the chemical properties of a substance are to subject the substance to a chemical change.

This provides a distinction between "chemical properties" and "physical properties." This distinction is not always clearly made in introductory chemistry courses.

The Learning Activity

1. **Predict** the products of the two reactions. **Explain** your predictions, using both balanced chemical equations and particle diagrams.

2. Obtain roughly 10 mL each of H$_2$SO$_4$ and Ba(OH)$_2$ solutions. Add 5 drops of universal indicator to each solution, and determine the pH of each solution.

3. Using a dropper, add the H$_2$SO$_4$ to the Ba(OH)$_2$ until you achieve a neutral green colour.

4. Obtain 10 mL of Ca(OH)$_2$ solution, and add 5 drops of universal indicator.

5. Using a straw, bubble your breath through the solution until you achieve a neutral green colour. Continue bubbling until the solution reaches pH 5.

4. **Observe** and **Explain** the results of both procedures

Equipment, Preparation and Resources

Dilute solutions of Ba(OH)$_2$ and H$_2$SO$_4$ ≈0.05 M
Limewater, i.e. saturated Ca(OH)$_2$
Universal indicator
test tubes, 100 mL beaker, drinking straws, disposable droppers,

Conductivity tester and burette if you wish to demonstrate the loss of conductivity in the barium sulfate reaction.

Categories:
Knowledge and Understanding:
Thinking and Inquiry:
Communication:
Applications / Connections:

Assessment and Evaluation
quality of explanation

clarity of drawings, equations, arguments

© Ross Lattner Publishing   **25**   www.rosslattner.com

Explaining Chemical Processes

Consider a simple grammatical rule. A concept map of the rule might resemble this linear structure.

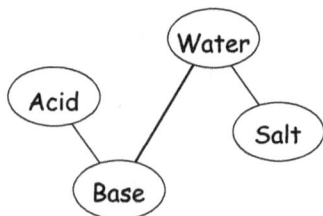

A second grammatical rule would have a similar structure:

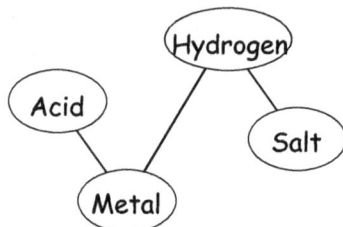

Ultimately, we would like our students to abandon linear rules and adopt a rich conceptual network. .

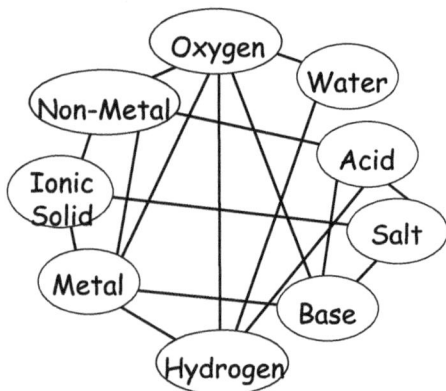

A student familiar with this network can navigate among many points of view

Lab 3.3: Reactions of Acids and Metals

Learning Expectations : *CP 1.05*: explain the interrelationships among metals and non-metals, acidic and basic oxides, and acids, bases, and salts; *CP 2.08*: represent simple chemical reactions using molecular models, word equations, and balanced chemical equations; *CP 2.09*: compare theoretical and empirical values and account for discrepancies when investigating conservation of mass; *CP 2.13*: conduct appropriate chemical tests to identify common gases.

Pedagogical Issues The relationships among metals, non-metals, acids, bases and salts form a conceptual network. For example, we teachers frequently write:

Acid + Base $\Longrightarrow$ Water + a Salt

but seldom write this equally important relationship:

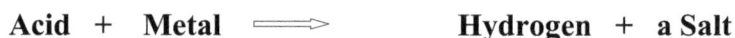

Acid + Metal $\Longrightarrow$ Hydrogen + a Salt

Conceptually, when we use simple grammatically structured rules, we ask the student to move along a linear conceptual track, a linear concept map as it were. It can easily be the case, for instance, that a student could learn both of the grammatically structured rules above, and not recognize that the acid and the salt could be the same in both cases. The more paths that we open up for students to move about on the field of chemistry, the more knowledge connections that the student can construct.

Science Issues This lab involves a 1.0 M solution of hydrochloric acid. While many teachers feel that this solution could safely be handled by their students under supervision, others may choose do perform this lab as a demonstration.

The aluminum metal is added to the cool acid solution. As the temperature of the acid is raised, the reaction proceeds quite efficiently.

When evaporated at low temperatures $90\,°C$, this reaction produces hydrates. If the evaporated residue is heated to, say, $120\,°C$, some hydrates will further dry to the anhydrous form. Are your students able to consider the presence of water in the crystalline solid?

Reactions of Acids and Bases

The safe way to test for hydrogen is to collect some of the gas in an inverted test tube, and bring it to a Bunsen burner at least 50 cm away.

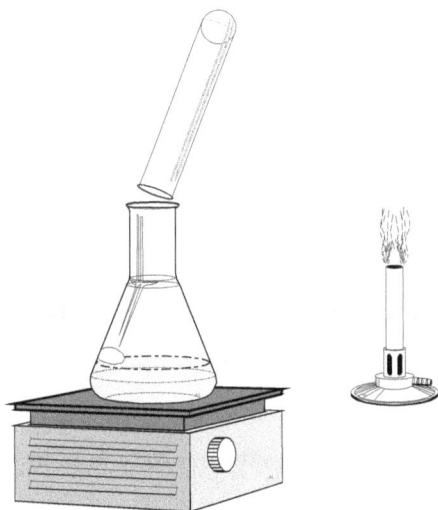

This lab is easily extendible to other metals and acids. It could become the basis of an inquiry.

The Learning Activity

1. The students **predict** the mass of $AlCl_3$ produced from 0.54 g of Al, and **explain** their predictions. Diagrams / sentences.

2. Obtain a clean, dry 250 mL flask, and write your name on the white spot with pencil. Weigh, and record the mass.

3. *Tare* the empty flask (zero the balance with the flask on it). Add bits of Al foil until the mass of Al is very nearly 0.54 g .

4. Add 100 mL of 1.0 M HCl. This is a slight excess.

5. Place the flask on a wire gauze, on a hot plate. Warm the HCl gently until the reaction begins. *Turn off the hot plate.* Collect some of the gas in an inverted test tube, and bring it to a Bunsen burner as shown.

6. When the reaction is over, place the flask in an oven at just under $100°C$, until the liquid has evaporated.

7. Measure the mass of the cool, dry flask and residue. Examine the residue. Write your **observations** and **explain** the results

For the metals, you may substitute 0.24 g Mg or 0.40 g Ca

For the acid, you may substitute 50 mL of 1.0 M H_2SO_4

These will produce different products, with different masses

Equipment, Preparation and Resources   for each group:

250 mL flask, test tube, Bunsen burner.
100 mL 1.0 M HCl, Al foil

You also need an oven, capable of reaching 150 °C

Categories:	Assessment and Evaluation
Knowledge and Understanding:	quality of predictions, explanations
Thinking and Inquiry:	explanation of any discrepancies
Communication:	quality of diagrams, sentences, argument, etc.
Applications / Connections:	extensions into other areas, e.g., hydrates, other metals.

When a student is learning, the place for a teacher to intervene is a very narrow zone.

If the student can already do the task, he or she doesn't need us.

If the understanding is beyond the student's ability, our intervention is just so much wasted noise.

The narrow zone for teacher intervention is the place where the student comprehends the general direction of the goal, but is having difficulty reaching it.

In this exercise, the students will be able to solve the simplest examples quickly. On the more difficult examples, some students may need help.

The student can gain a sense of understanding and control if we direct him or her to draw Dalton pictures of the reaction.

The pictures can provide a student with "another way of seeing" a problem

Activity 3.4: Balancing Neutralization Reactions
Activity 3.5: Balancing Reactions of Metals and Acids

Learning Expectations : *CP* **1.01**: recognize the relationships among chemical formulae, composition, and names; *CP* **1.02**: explain, using the law of conservation of mass and atomic theory, the rationale for balancing equations; *CP* **1.08**: name and write the formulae of common ionic and molecular compounds

Pedagogical Issues We teachers have become so used to representing chemical changes in terms of balanced chemical equations, that we run the risk of thinking that chemical equations *are* the same as chemical changes.

Balanced chemical equations are book-keeping devices. They keep track of a molecular balance sheet, so that we don't violate the law of conservation of matter. As book-keeping devices, they are useful, but not absolutely true.

When kids begin to learn how to balance equations, they appear to have three problems:

1. They do not recognize the atomic symbols, and comprehend how to read them. The reason we make all atomic symbols consist of one Capital letter, followed by small letters, is to help kids (and adults) recognize where one symbol leaves off and the next one begins.

2. Kids don't understand the meaning of the subscripts. In a chemical formula such as H_2SO_4, novices sometimes find it difficult to understand that the numeral refers only to the object immediately to its left. This becomes more difficult when we have brackets as in $Ca(OH)_2$ Kids with learning difficulties may find this especially confusing.

3. Kids don't understand the meaning of the coefficients. Beginners almost always confuse the roles of the coefficients and subscripts.

Science Issues as the students complete the balanced equations, how can they know if they are right?

The students have already completed the exercise naming acids and bases. In those exercises, the correct name and formula for each acid and base is provided in a number of repetitions.

It is impossible to discuss chemical formulas without some knowledge of the electron configuration of each element. Students should be provided copies of the periodic tables at the back of the lab manual. On those tables, the electron configurations are provided for reference, and the ability of each atom to gain or lose electrons is also provided.

Finally, any of these reactions could conceivably be demonstrated in the classroom. Consult your school's safety policies before attempting any demonstrations

The Learning Activity

Balance the equations as given. The students may need to review the process. Always use Dalton diagrams when you demonstrate how you balance equations.

Equipment, Preparation and Resources

Photocopy pages from the student manual

Pens, pencils, etc.

Categories:
Knowledge and Understanding:
Thinking and Inquiry:
Communication:
Applications / Connections:

Assessment and Evaluation
accuracy of the balanced equations.

Spelling of names, accuracy of formulas

Explaining Chemical Processes

Lab 3.6: Rate of Reaction of Metal with Acid

How can a student measure reaction rate? The simplest approach is to have the kids measure the time between the initiation of a reaction and some decisive event, such as a colour change or popping the top off a film cannister

This provides a good qualitative comparison of rate, but does not provide much detail about the processes occurring along the way.

One of the most important features of chemical processes is that rate decreases as chemicals are used up. This lab provides some data on the relationship between the amount of acid, measured as pH, and the rate of production of hydrogen gas.

Learning Expectations : CP 1.04: describe and explain qualitatively how factors such as energy, concentration, and surface area can affect rates of chemical reactions; **CP 2.12:** design an experiment to determine qualitatively the factors that influence chemical reactions.

Pedagogical Issues Rate is known by human beings as the direct experience of time. For example, when children first begin to talk about rates of consumption, they are concerned with the time it takes to gobble a cookie, not so much with the amount of cookie eaten. In fact, the amount of cookie is simply seen as a complicating factor, not as the main event.

For scientists, however, it is important that we do not equate rate with time. The rate of a chemical reaction is not just "how long does it take?" Chemical rate is also concerned with "how much is being used up?" or "how much is being produced?"

How can we convey this to beginning students, without using the concept of the mole? This lab has two observable events for students to consider. First, we can observe the pH as a colour change in universal indicator or a pH meter. Second, we can observe the volume of hydrogen gas produced.

Both of these quantities change over the time of the reaction. Students familiar with the pH scale should note that there is not nearly as much H^+ in the flask at pH 3 as there was at pH 1. The student could make a meaningful correspondence between H+ and production of hydrogen.

Science Issues The reaction in the flask is one that students have studied before in this class. They are already familiar with the reactants, products, and the properties of metals and non-metals

Each test will require about 20 mL of 0.10 M HCl. This is a total of 2 millimoles of HCl. The hydrochloric acid is the limiting reagent in this lab: it will be completely consumed..

A typical piece of Calcium turning has a mass between 0.1 and 0.4 g, or about to 2.5 to 10 mmols, more than enough to consume the HCl.

Reactions of Acids and Bases

This lab lends itself to exploration of a variety of factors that might affect rate.

Students can dilute the acid by adding 20 ml of water to each sample.

The temperature can easily be modified by placing the flask in a warm or cold water bath.

Solid Mg metal and powdered Mg can be substituted for the Ca metal

Sulfuric acid or acetic acid can be substituted for the hydrochloric.

When the HCl has completely reacted with Calcium, it will produce 1 millimole of H_2 gas, or 22.4 mL @STP. This volume of gas is easily measured in the graduated cylinder.

The calcium metal is usually sold as metal turnings, and the pieces are not completely uniform. You should select pieces as similar as possible for this experiment. Furthermore, the calcium metal is oxidized on the surface. It can be oxidized deeply into the metal, especially if not used for a period of years. Fresh calcium works best for this lab.

The Learning Activity

1. Students will start with 0.10 M HCl, about pH ≈ 1. They will **predict** how the pH will change, and how the speed of the reaction will change as the pH changes, and **explain** their prediction.
2. Assemble the apparatus shown at right. Your teacher will provide 20 mL of 0.10 M HCl, and a piece of Ca metal. Using forceps, place the Ca inside the flask, and replace the stopper. Start timing when the reaction begins.
3. **Observe** and record the pH and volume of H_2 gas collected every 30 s. **Explain** your results

Equipment, Preparation and Resources

Samples of Ca metal, selected for approximate uniformity.

0.10 M HCl, 20 mL for each experiment.

You may need to photocopy extra sheets of the graph paper in the student exercises to accommodate additional experiments.

Categories:	Assessment and Evaluation
Knowledge and Understanding:	quality of answers to questions for later
Thinking and Inquiry:	quality of design for further exploration
Communication:	quality of diagrams, graphs, written arguments
Applications / Connections:	

Lab 3.7: Reaction of Acids with Carbonates

A researchable question for the teacher. How do students come to terms with this set of reactions?

Do more of them make sense of the essential change using the balanced symbolic equation? Or do more of them make sense of the Dalton pictures to discern the nature of the decomposition?

This kind of question can make teaching much more interesting, and provide support for your own professional satisfaction and growth.

Learning Expectations : *CP 1.08*: name and write the formulae of common ionic and molecular compounds, using a periodic table and an IUPAC table of ions; *CP 2.13*: conduct appropriate chemical tests to identify common gases.

Pedagogical Issues The final lab in this unit, provides one more opportunity to make representation of a variety of chemical reactions in the laboratory.

In this case, we end up where we started, with the carbonate ion. The acid turns the carbonate ion to carbonic acid, which rapidly decomposes into carbon dioxide and water. There is a remarkable number of circumstances in which this reaction can occur.

One of the cognitive issues in this lab exercise concerns the student's ability to recognize patterns of continuity in the representational system, and relate them to patterns of behaviour in the lab. The carbonate ion appears in every equation, and decomposes into carbon dioxide and water. The hydrogen ion appears on every acid, and is essential in the decomposition of carbonate. This is not at all evident to the first time observer.

Science Issues You need not have strong acids for this lab, but it is important to provide enough for several trials.

Reactions of Acids and Bases

The test for carbon dioxide surfaces again.

The gas can be collected as it "pours" out of the acid plus carbonate mixture. A good time to remark upon the mass of the carbon dioxide molecule.

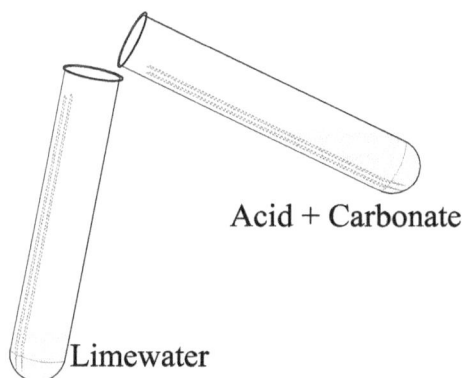

Acid + Carbonate

Limewater

The gas collected in the limewater test tube rapidly reacts to provide the milky white precipitate.

The Learning Activity

Students will be given 2 acids and 2 carbonates:

sulfuric acid	H_2SO_4	0.10 M	× 100 mL
hydrochloric acid	HCl	0.10 M	× 100 mL
calcium carbonate	$CaCO_3$	several small stones	
sodium carbonate	Na_2CO_3	5 g	

In four separate experiments, they will combine each acid with each carbonate.

1. **Predict** the products formed with each combination. **Explain** your predictions, using diagrams, balanced equations, and sentences.
2. Do the experiment. Test the gas produce. Is it CO_2 ? **Observe** and **Explain** your observations.

Equipment, Preparation and Resources

For each pair fo students:

100 mL of 0.10 M HCl
100 mL of 0.10 M H_2SO_4
several lumps of calcium carbonate, marble chips or chalk
a spoonful of sodium carbonate

25 mL of limewater

8 test tubes

test tube rack

Categories:

Knowledge and Understanding:	
Thinking and Inquiry:	
Communication:	
Applications / Connections:	

Assessment and Evaluation

quality of representations, predictions and explanations
quality of arguments
quality of diagrams, equations, sentences

Student Exercises

Explaining Chemical Processes

Knowledge and Understanding

Two theories are emphasized in this unit. Dalton's theory of chemical change relates observable chemical changes to the invisible rearrangements of atoms into new molecules. Arrhenius' theory provides a useful model of the behaviour of strong acids and bases in aqueous solutions. You can use these theories to predict and explain a large number of events in the world around you. Neither theory is big enough to explain everything. You will eventually need to learn newer, broader theories that explain even more things.

You will also learn to use four parallel representations to depict chemical change. You will use Dalton pictures, balanced chemical equations, word equations, and mass equations.

Your growing knowledge and understanding will be probed at regular intervals in the Grade Ten Daily quizzes. Study these as you go through the exercises, so that you can do your best when they are assigned.

Inquiry and Thinking

We will use the PEOE cycle for most labs and activities. You are expected to frame a question, provide your best prediction, and explain your thinking, using both sentences and diagrams.

Communication

The quality of your arguments is the most important aspect of communication in this chapter. Your arguments consist of sentences, organized into paragraphs, and supported by diagrams or other representations.

Each sentence should be clear and to the point. When you are learning new concepts, you will find it best to limit your sentences to two concepts linked together to make a reasonable claim. If you need to relate more than two concepts, add a new sentence.

Applications, Connections and Extensions

Every exercise in this book is designed to support you as you learn appropriate theories and apply them to problems. In the labs, you demonstrate your understanding of a theory only by applying the theory. In the quizzes and projects, you are invited to make further connections and extensions of your learning.

Explaining Chemical Processes

Introduction: Three Ideas of Chemical Change

In this unit, we will explore how one kind of matter can change into another kind of matter. Such changes, which occur all around us, are called chemical changes. This unit consists of two main ideas:

1. **Dalton's Theory of Chemical Change** Consisting of six simple statements, this theory summarizes and integrates important features of Dalton's theory and the kinetic molecular theory to support the new ideas you will learn in this unit. Whenever a chemical change takes place, some things are changed, but other things remain exactly the same. Consider this chemical change. Two chemicals which smell like beer and rancid butter combine and make the stuff that flavours apple candies and drinks.

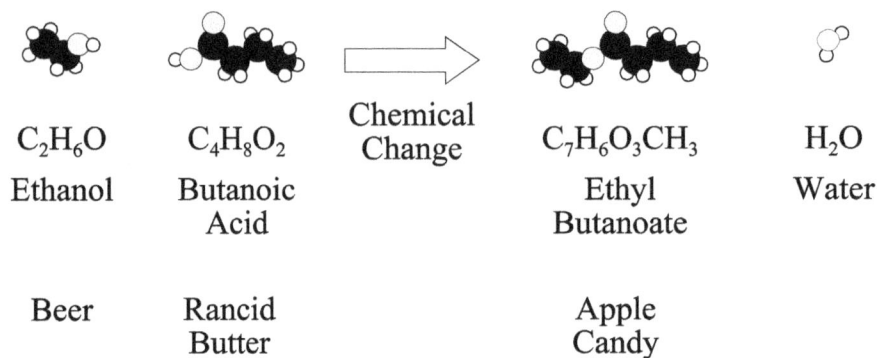

C_2H_6O	$C_4H_8O_2$	Chemical Change	$C_7H_6O_3CH_3$	H_2O
Ethanol	Butanoic Acid		Ethyl Butanoate	Water
Beer	Rancid Butter		Apple Candy	

1. **All of the atoms present before a chemical change are still present after the change.** We can use this knowledge to "count" atoms before and after chemical changes, and to make predictions about the kinds of products that might be formed in a chemical change.

2. **The total amount of mass (matter) is the same before and after a chemical change.** We are so certain of this idea, that even if the water molecules "got away on us," we could still measure the mass of the missing water molecules.

3. **During a chemical change, atoms are rearranged into new groups.** In the example above, the reactant atoms rearrange to make different groups of atoms in the products.

4. **During a chemical change, new particles of new substances are formed.** A new arrangement of atoms is a new substance. The flavours on the left are awful. When reacted, the new particles have a very pleasant flavour.

5. **During a chemical change, old chemical bonds are broken, and new chemical bonds formed.** Which bonds were broken in the example above? Which new bonds were formed?

6. **Chemical bonds are electrical forces of attraction between valence electrons and two nuclei.** All chemical changes are the result of changing electrical forces between atoms.

2. **Arrhenius' Theory of Acids and Bases** explained many characteristics of acids and bases when they were dissolved in water.

    1.   **Acids are compounds which contain hydrogen ions (H^+).**

    2.   **Bases are compounds which contain hydroxide ions (OH^-).**

    3.   **Non-metal oxides form acidic solutions in water.**

    4.   **Metal oxides form basic solutions in water.**

    5.   **An acid plus a base react to produce neutral water, plus a salt.**

    6.   **A neutral solution contains equal numbers of hydrogen and hydroxide ions.**

You will test some of the statements about acids and bases to see if they appear to be true. Then you'll use these statements to figure out what's going on in some other experiments.

In addition to these two theories, you must represent chemical change in four different ways:

	Reactants	Chemical Change	Products

Pictorial shows how the reactant atoms are bonded together, and how they rearrange into new product molecules.

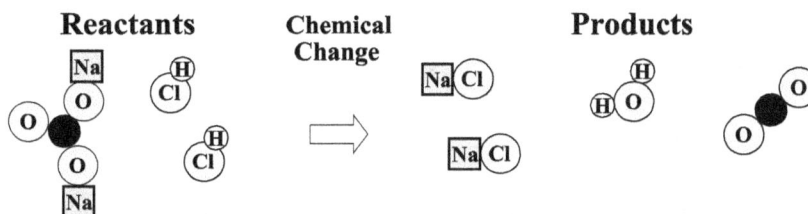

Chemical Equations summarize the pictorial with chemical symbols.

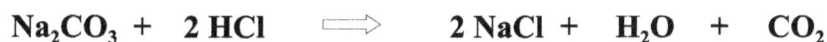

$$Na_2CO_3 \ + \ 2\,HCl \ \Longrightarrow \ 2\,NaCl \ + \ H_2O \ + \ CO_2$$

Chemical Names allow us to talk about reactions.

Sodium Carbonate + Hydrochloric Acid $\Longrightarrow$ Sodium Chloride + Water + Carbon Dioxide

Masses allow us to measure how much reactant and product.

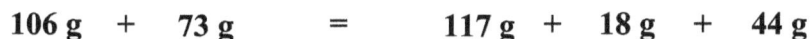

$$106\,g \ + \ 73\,g \ = \ 117\,g \ + \ 18\,g \ + \ 44\,g$$

One of the tasks of every scientist is to choose which kinds of representations are most useful for each new question. Which representation should you use? It depends upon the situation. You will learn by experience which representations are most helpful.

In these exercises, the question must be answered in *complete sentences*. One sentence is one thought. A single word is simply not enough.

Explaining Chemical Processes

Lab 1.1: Decomposition of Ammonium Carbonate

Do you Remember? List the 6 propositions of Dalton's Theory of Chemical Change

1. _____
2. _____
3. _____
4. _____
5. _____
6. _____

What's The Question? We know that when temperature of ammonium carbonate is raised, its particles will speed up (the Kinetic Molecular Theory).
Will the particles of ammonium carbonate move fast enough for some of them to break free and leave the solid?

What Are We Doing?

1. **Predict** which particles, if any, will break free of the ammonium carbonate molecule. **Explain** your prediction using words and pictures.

2. Place 0.96 g of ammonium carbonate in a test tube. Clamp as shown. Heat gently.

3. Test the gases at the mouth of the test tube. See the *"What are we thinking about?"* section at right.

4. Record what you **Observe**. **Explain** what happened, using pictures and sentences.

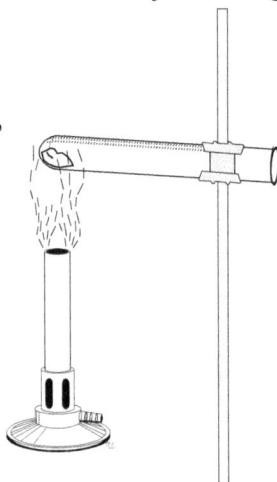

What Are We Thinking About?

Caution: irritating gases. Do not inhale. Smell by wafting some gases toward your nose. *Never take your goggles off during this lab: even if you finish, others may still be working.*

- At room temperature, air and water particles are moving at about 450 m/s, faster than sound.

- **Test for water:** dry cobalt chloride paper is blue. In the presence of water, it turns pink.

- **Test for ammonia:** dampen a strip of red litmus paper. In the presence of ammonia, it turns blue.

Questions For Later...

1. What chemical bonds are broken in this reaction? What new chemical bonds are formed?

2. What is the difference between *ammonia gas* and the *ammonium ion*?

3. If you had exactly 0.96 g of $(NH_4)_2CO_3$, how many grams of each product would be produced?

Decomposition and Synthesis

Name:

Date:

K I C A

Focus Question: Write the question that you are trying to answer.

Before the experiment...
Complete this diagram to show four different representations of the chemical change that you predict.

In your pictures, chemical formulae, and masses, you must show that no matter has been created or destroyed.

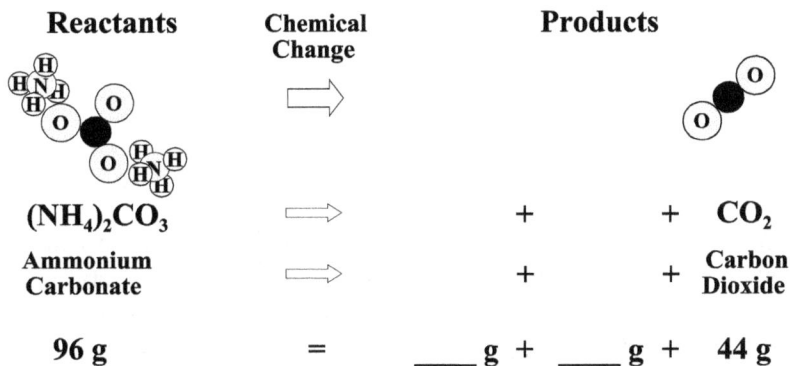

Reactants

H H N H H O O O O H H N H

$(NH_4)_2CO_3$

Ammonium Carbonate

96 g

Chemical Change

⇒

⇒

⇒

=

Products

O O O

+   + CO_2

+   +   **Carbon Dioxide**

___ g + ___ g + **44 g**

---

1   *Predict*

2   *Explain*

3   *Observe*

4   *Explain*

Explaining Chemical Processes

Lab 1.2: Decomposition of Copper (II) Sulphate Pentahydrate

Do you Remember? **List the 6 propositions of Dalton's Theory of Chemical Change**

1. _____
2. _____
3. _____
4. _____
5. _____
6. _____

What's The Question? Many compounds include some water molecules in their crystals. These water molecules occupy spaces between the positive metal ions and the negative ions.
When we heat such a substance, will we drive off its water of crystallization or cause some other kind of reaction? If 2.50 g of copper sulphate is heated, what mass of water would be driven off?

What Are We Doing?

Predict the mass of water that will break free of the $CuSO_4 • 5 H_2O$ crystals. **Explain** your prediction using words and pictures.

1. Find the mass of a clean dry test tube.
2. Place 2.50 g of $CuSO_4 • 5 H_2O$ in the test tube. Clamp as shown. Heat gently at first, then more strongly.
3. Heat the entire test tube just enough to drive off any condensed water.
4. Let the test tube cool to room temperature. Find the mass of the test tube + product.
5. Add 2 - 3 drops of water to the cool product. Note any temperature change.

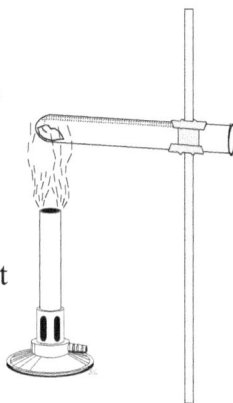

Record what you **Observe**, and **Explain** it.

What Are We Thinking About?

Caution: Hot glass burn hazard. Let the apparatus cool at least 15 min. *Never take your goggles off during this lab: even if you finish, others may still be working.*

- A *decomposition* is a chemical reaction in which a large molecule is broken into simpler molecules.

- Anhydrous copper sulphate is a white powder containing no water.

Questions For Later...

1. What mass of water was driven off?

2. What is the product left behind in the test tube?

3. What happened when you added a drop or two of water to the anhydrous product?

Decomposition and Synthesis

Focus Question: Write the question that you are trying to answer.

Before the experiment...
Complete this diagram to show four different representations of the chemical change you predict.

In your pictures, chemical formulae, and masses, you must show that no matter has been created or destroyed.

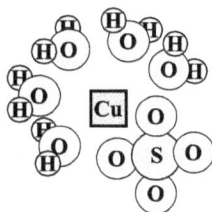

_____ CuSO$_4$ • 5 H$_2$O
**Copper
Sulphate
Pentahydrate**

250 g

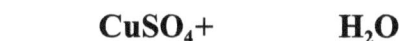 _____ CuSO$_4$+ _____ H$_2$O

 +

= _____ g + _____ g

1	***Predict*** the mass of water driven off when 2.50 g of CuSO$_4$ • 5 H$_2$O is heated.	2	***Explain*** your prediction, in both diagrams and words.

3	***Observe*** the results of the heating. Make records of everything that you see.	4	***Explain*** your observations, using both pictures and sentences.

_____ mass of clean dry test tube
_____ mass of test tube + CuSO$_4$ • 5 H$_2$O
_____ mass of CuSO$_4$ • 5 H$_2$O
_____ mass of test tube + CuSO$_4$
_____ mass of CuSO$_4$
_____ mass of H$_2$O driven off

Explaining Chemical Processes

Lab 1.3: Using Iron to Find % Oxygen in Air

Do you Remember? **List the 6 propositions of Dalton's Theory of Chemical Change**

1. _____
2. _____
3. _____
4. _____
5. _____
6. _____

What's The Question? We know that the atmosphere contains oxygen. That's the gas that we need to breathe in order to live. Is air 100% oxygen? 50% oxygen? Or only 10% oxygen? *What % of the earth's atmosphere is oxygen? What are the other gases?*

What Are We Doing?

1. **Predict** the percentage of the earth's atmosphere that is oxygen, O_2. **Explain** your prediction in diagrams and sentences.

2. Obtain a tuft of steel wool about the size of a large grape. Clean it first in alcohol, then in dilute hydrochloric acid to remove any grease and oxide film. Rinse with water.

3. Fluff up the steel wool. Push it gently to the bottom of the cylinder. Turn the cylinder upside down, and trap exactly 100 mL of air inside.

4. Leave it for several days. The water level will rise as the oxygen reacts with the iron. Read the final volume of gases.

5. **Observe** all changes, including appearances of the iron, as well as the volume of gas. **Explain** your observations in diagrams and sentences.

What Are We Thinking About?

Oxygen will strongly attract the electrons from iron, forming iron oxide. This reaction will completely react the oxygen in air. If air is 60% oxygen, how much of the gas will be consumed in the reaction?

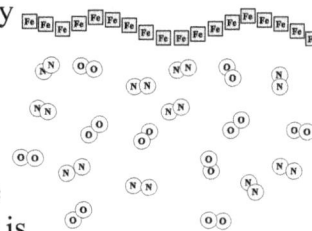

The formation of acids or bases can be a sign of a chemical change. To test for an acid or base, use ***pH test paper***. This paper is green in the presence of neutral solutions like water. It will turn yellow, then orange and finally red in the presence of acids. It will turn green-blue, and finally dark blue in the presence of bases.

Questions For Later...

1. What volume of gas was consumed? What is the % oxygen in air?

2. What evidence is there that iron and oxygen combined to form iron oxide?

Decomposition and Synthesis

Focus Question: Write the question that you are trying to answer.

Before the experiment...
Complete this diagram to show four different representations of the chemical change you predict.

In your pictures, chemical formulae, and masses, you must show that no matter has been created or destroyed.

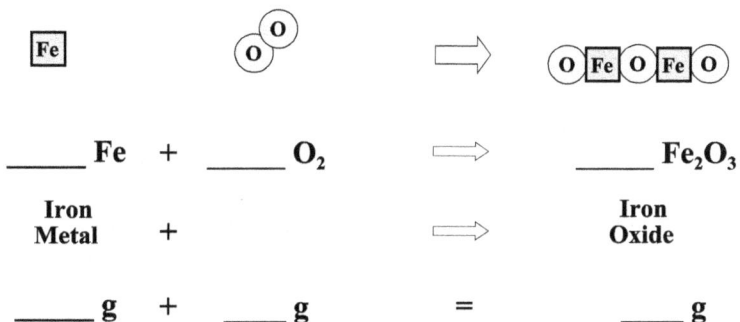

_____ Fe + _____ O_2 $\Rightarrow$ _____ Fe_2O_3

Iron Metal + $\Rightarrow$ **Iron Oxide**

_____ g + _____ g = _____ g

1   **Predict** You will start with 100 mL of air trapped in a graduated cylinder. What will be the volume of gas left in the cylinder after the oxygen has reacted with the iron?

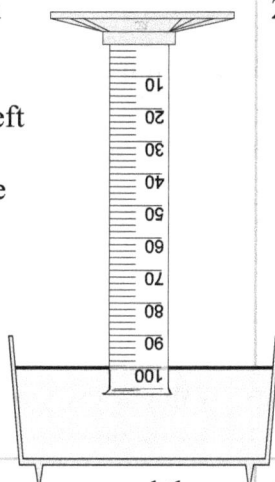

2   **Explain** your prediction, using both diagrams and sentences.

3   **Observe** the changes in appearance, and the results of any tests. Include all measurements:

_____ Volume of trapped air before reaction
_____ Volume of gas left unreacted
_____ Volume of oxygen that reacted
_____ % oxygen in air

4   **Explain** your observations, using both diagrams and sentences.

Lab 1.4: Synthesis of Magnesium Oxide

Do you Remember? **List the 6 propositions of Dalton's Theory of Chemical Change**

1. _____
2. _____
3. _____
4. _____
5. _____
6. _____

What's The Question? Magnesium metal, like iron, is very reactive with oxygen. It can form a compound, magnesium oxide which is quite different from magnesium metal.

What mass of magnesium oxide, MgO, will form from 0.24 g of magnesium metal?

What Are We Doing?

1. Weigh a clean dry crucible and lid.
2. Cut a piece of Mg ribbon with a mass of 0.24 g. Coil it into a spiral, and put it into a crucible.
3. Measure the mass of the crucible + lid + magnesium..

Predict mass of MgO that can be formed from 0.24 g of Mg metal.
Explain your prediction in diagrams and sentences.

4. Put a lid on the crucible at a slight angle. Gently heat the crucible for 5 min. Then heat to red heat for about 10 min.
5. Carefully open the lid about 1 mm with tongs. If smoke escapes, close the lid. Heat another 5 min, gradually removing the lid.
6. Allow the crucible to cool for 15 min.
7. Weigh the crucible + lid + product MgO.
8. Examine the product.

Observe all changes, including appearances of the magnesium and the product. **Explain** your observations in diagrams and sentences.

What Are We Thinking About?

Caution: hot objects and danger of burns. Allow everything to cool for 15 min before attempting to measure the mass. Use tongs to handle the hot lid.

When a metal burns in oxygen, does it increase in mass or decrease in mass?

Questions For Later...
1. What mass of MgO did your reaction produce? Was your prediction correct? Explain.

2. A student lifted the lid too early in step 5. A puff of white smoke escaped. *Does the smoke have mass? Explain.*

Decomposition and Synthesis

Name:

Date:

Focus Question: Write the question that you are trying to answer.

Before the experiment...
Complete this diagram to show four different representations of the chemical change you predict.

In your pictures, chemical formulae, and masses, you must show that no matter has been created or destroyed.

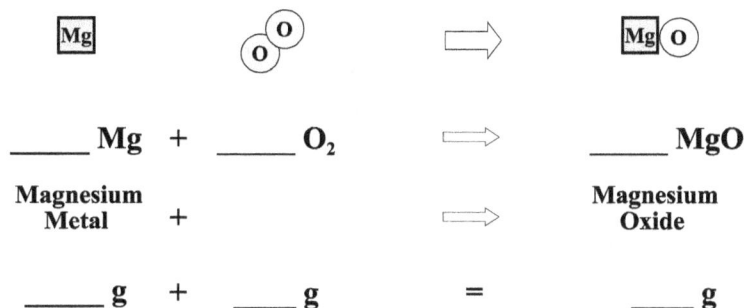

| Mg | O O | ⟹ | Mg O |

_____ Mg + _____ O_2 ⟹   _____ MgO

Magnesium Metal +   ⟹   **Magnesium Oxide**

_____ g + _____ g   =   _____ g

---

1 **Predict** the mass of MgO produced when 0.24 g of Mg are burned in the crucible.

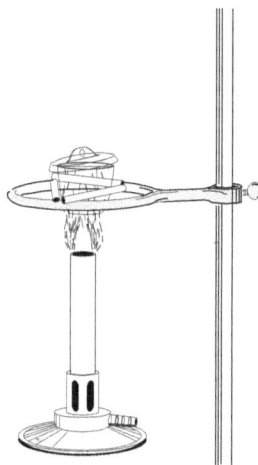

2 **Explain** your prediction, using both diagrams and sentences.

---

3 **Observe** the changes in appearance, and the results of any tests. Include all measurements:

_____ Mass of clean dry crucible
_____ Mass of crucible + Mg
_____ Mass of Mg (should be about 0.24 g)
_____ Mass of crucible + MgO
_____ Mass of MgO

4 **Explain** your observations, using both diagrams and sentences.

Quiz 1.5: Decomposition and Synthesis

1

Draw Pictures

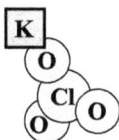

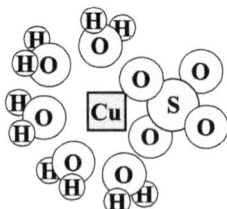

Mg   O O   ⟹   Mg O

Balance the Equation   _____ Mg + _____ O_2 ⟹   _____ MgO

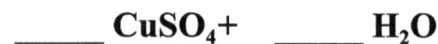

Write the Names   **Magnesium Metal** +   ⟹   **Magnesium Oxide**

Write the Masses   _____ g + _____ g   =   _____ g

Date: _____   / 5

2

Draw Pictures

K O Cl O O   ⟹   K Cl   O O

Balance the Equation   _____ $KClO_3$ ⟹   _____ KCl + _____ O_2

Write the Names   **Potassium Chlorate**   ⟹   **Potassium Chloride** +

Write the Masses   _____ g   =   _____ g + _____ g

Date: _____   / 5

3

Draw Pictures

H H O H H O H H O Cu S O O O O H O H H O H   ⟹

Balance the Equation   _____ $CuSO_4 \cdot 5\,H_2O$ ⟹   _____ $CuSO_4$ + _____ H_2O

Write the Names   **Copper Sulphate Pentahydrate**   ⟹   +

Write the Masses   **250 g**   =   _____ g + _____ g

Date: _____   / 5

Quiz 1.5: Decomposition and Synthesis   **Name:**

4

Draw Pictures

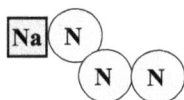

Balance the Equation   _____ Fe + _____ O_2 ⟹   _____ Fe_2O_3

Write the Names   **Iron Metal** +   ⟹   **Iron Oxide**

Write the Masses   _____ g + _____ g   =   _____ g

Date: / 5

5

Draw Pictures   ⟹

Balance the Equation   _____ NaN_3 ⟹   _____ Na + _____ N_2

Write the Names   **Sodium Azide (in airbags)** ⟹ **Sodium Metal** +

Write the Masses   _____ g   =   _____ g + _____ g

Date: / 5

6

Draw Pictures   ⟹

Balance the Equation   _____ N_2 + _____ H_2 ⟹   _____ NH_3

Write the Names   **Nitrogen** + _____   ⟹   **Ammonia**

Write the Masses   _____ g + _____ g   =   _____ g

Date: / 5

All the news that's fit to print... and then some

The Grade Ten Daily

Quiz 1.5: Decomposition and Synthesis   Name:

7 Complete the periodic table below as shown in the example. In each cell of the table, print neatly the core charge, the atomic number, the element symbol, and draw the valence electrons.

2.1							+4 14 Si	--
1.0	1.5	2.0	2.5	3.0	3.5	4.0		--
0.9	1.2	1.5	1.8	2.1	2.5	3.0		--
0.8	1.0	1.6	1.8	2.0	2.4	2.8		--

Date: / 5

The **electronegativity** of an element is *the ability of an element to grab and hold electrons from another atom.* The electronegativity is a number between 0 and 4, found in the top right corner of each cell above. The most electronegative element is fluorine, with ε = 4.0.

The **core charge** of an atom is the total electric charge of the nucleus plus the core electrons. It is given by the number in the centre of each atom.

The **valence electrons** occupy the outer regions of the atom, and define the **atomic radius**.

8 As you read the table from left to right, the core charge increases steadily. What effect does this have on the electronegativity? Why?

As you read the table from top to bottom, the radius of the atoms increases. What effect does this have on electronegativity? Why?

Date: / 5

9 In question 1, you worked with magnesium, oxygen, and magnesium oxide.

Is magnesium's core charge **greater** than, **less** than or **equal** to oxygen's core charge?

Is magnesium's radius **greater**, **less** or **equal** to oxygen's radius?

Is magnesium's electronegativity **g l e** to oxygen's electronegativity?

If oxygen and magnesium engaged in a tug of war for electrons, who would win, and who would lose? Explain, using core charge, radius and electronegativity.

Date: / 5

10 In question 2, you studied the decomposition of potassium chlorate. Circle the letter to:

Compare core charges: is K **g l e** Cl?
Compare radius: is K **g l e** Cl?
Compare electronegativity: is K **g l e** Cl?

If K and Cl engaged in a tug of war for electrons, who would win? Explain, using core charge, radius and electronegativity.

Date: / 5

11 One of the products of the decomposition of potassium chlorate is oxygen gas, O_2. The oxygen molecule consists of two oxygen atoms. Circle the letter to:

Compare core charges: is O **g l e** O?
Compare radius: is O **g l e** O?
Compare electronegativity: is O **g l e** O?

If O and O engaged in a tug of war for electrons, who would win? Explain, using core charge, radius and electronegativity.

Date: / 5

12 Potassium chlorate, $KClO_3$ is a compound in which chlorine and oxygen are combined. Circle the letter to:

Compare core charges: is Cl **g l e** O?
Compare radius: is Cl **g l e** O?
Compare electronegativity: is Cl **g l e** O?

If Cl and O engaged in a tug of war for electrons, who would win? Explain, using core charge, radius and electronegativity.

Date: / 5

Explaining Chemical Processes

Lab 2.1: Reactions of Non-metals and Oxygen

Do you Remember? **List the 6 propositions of Dalton's Theory of Chemical Change**

1. _____
2. _____
3. _____
4. _____
5. _____
6. _____

What's The Question? Carbon, sulfur and oxygen are all non-metals.

Will carbon oxide produce an acid or a base? Will sulfur oxide produce an acid or a base?

What Are We Doing?

1. ***Predict*** an answer, and ***Explain*** your prediction.
2. Fill two gas bottles with oxygen gas, and cover them with glass plates. Add 1 cm of water, plus 10 drops of bromthymol blue.
3. Place 4 - 6 small lumps of charcoal (carbon) in the deflagrating spoon, and heat it in a Bunsen flame until it is red hot. Gently lower the spoon into the oxygen. Do not touch the water! Raise and lower the spoon slowly to stir up the oxygen.
4. Observe the reaction for 2 min, then withdraw the spoon, close the gas bottle, and gently swirl the bromthymol blue solution.
5. Put a very small (rice-grain size) piece of sulfur in the spoon. Gently warm it in the Bunsen burner flame until it just catches fire. Quickly lower it into the second gas bottle of fresh oxygen. Do not touch the water! Raise and lower the spoon slowly to stir up the oxygen.
6. Observe the reaction for 2 min. Withdraw the spoon, close the gas bottle, and gently swirl the bromthymol blue solution.

What Are We Thinking About?

Caution: hot objects and danger of burns.

Caution: sulfur oxide is a noxious gas. Use small quantities of sulfur, about the size of a rice grain.

- When a non-metal burns in oxygen, does it increase in mass or decrease in mass?

- Are the non-metal oxides acidic or basic?

Questions For Later...

1. What evidence do you have that an acid or base was produced?

2. Sulfur dioxide is produced in mining, manufacturing, and in the use of sulfur bearing gasoline. What products would be formed if sulfur dioxide was present in rain clouds? Explain.

3. Suppose that a family burned 32 kg of sulfur in their gasoline in a year. What mass of sulfur dioxide would that family produce?

Metals, Non-Metals and pH

Name:

Date:

Focus Question: Write the question that you are trying to answer.

1 **Predict:** will carbon dioxide and sulfur dioxide produce acids or bases?	2 **Explain** your thinking.
3 **Observe** the reactions. Were your predictions correct?	4 **Explain**, using any new ideas that you gained from this experiment.

After the experiment...
Complete this diagram to show four different representations of the chemical change you observed.

____ S   +   ____ O_2 ⟹   ____ SO_2

Sulfur   +   _____   ⟹   Sulfur Dioxide

____ g   +   ____ g   =   ____ g

After the experiment...
Complete this diagram to show four different representations of the chemical change you observed.

____ C   +   ____ O_2 ⟹   ____ CO_2

Carbon   +   _____   ⟹   _____

____ g   +   ____ g   =   ____ g

In each case, the oxide reacts with water to form an acid.

____ CO_2 +   ____ H_2O ⟹   ____ H_2CO_3

____ SO_2 +   ____ H_2O ⟹   ____ H_2SO_3

Explaining Chemical Processes

Activity 2.2: What Makes Acids Acidic?
Do you Remember? **List 6 propositions of Arrhenius' theory of Acids and Bases**

1. _____
2. _____
3. _____
4. _____
5. _____
6. _____

What's The Question? We observed that non-metal oxides formed acidic solutions when they dissolved in water. *Why? What is it about non-metals that makes them become acidic?*

What Are We Doing?
Each of the molecules on the following page is an acid that can be made by dissolving an oxide of the related non-metal in water.

1. Write the core charge in the core of each atom

2. Determine the electronegativity (ε) of each atom in the acid molecule, using the core - valence - radius table on page 48. Write the ε near each atom.

3. In the tug-of-war for electrons, who is "the big winner"? Which atom will attract electrons most strongly?

4. In the tug-of-war for electrons, who is "the big loser"? Which atom will lose its electrons to the others?

5. If "the big loser" in each case loses its electrons, what does it become?

What Are We Thinking About?
- Each atom in the periodic table has a characteristic core charge, radius, and set of valence electrons.

- Electronegativity is *the ability of an atom to grab and hold another atom's electrons.*

- Chemists believe that molecules are groups of atoms held together with chemical bonds. The bonds are believed to consist of pairs of electrons that are attracted to both of the atoms. The diagrams that follow are one kind of diagram that scientists use to describe chemical bonding.

Questions For Later...
1. The metalloids are the elements (B, Si, As, Te) that lie between metals and non-metals. All of them have ε ≈2.0. Which element in the acids you have considered has ε closest to 2.0?

2. Must *all* of the non-metal oxides form acidic solutions? Explain your answer.

Metals, Non-Metals and pH

Name:
Date:

1   Carbonic acid H_2CO_3

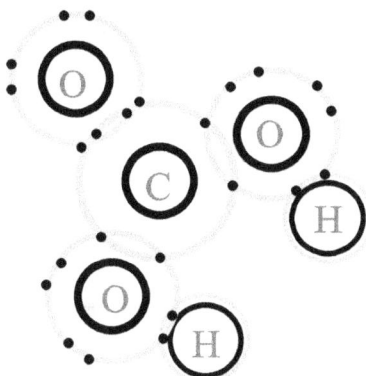

2   Sulfurous acid H_2SO_3

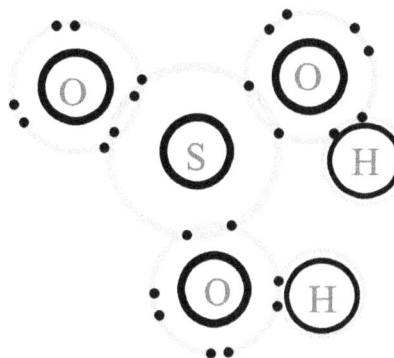

3   Nitric acid HNO_3

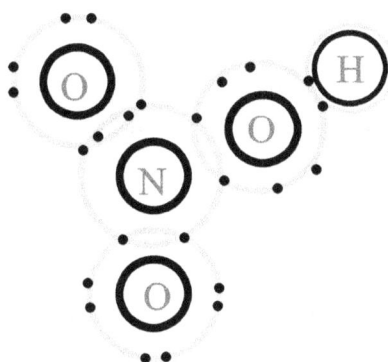

4   Sulfuric acid H_2SO_4

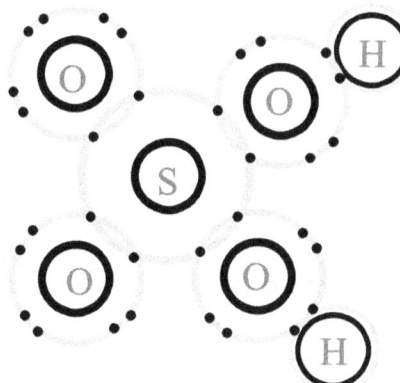

5   Acetic acid CH_3COOH

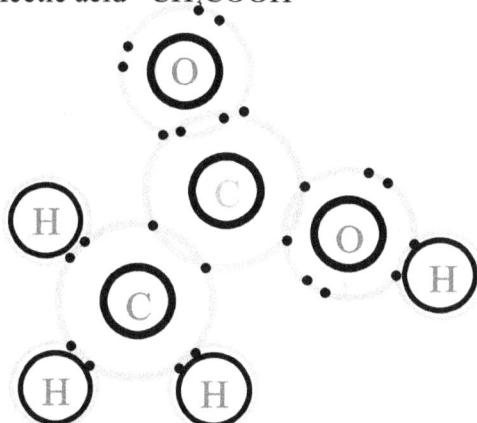

6   Hydrochloric acid HCl

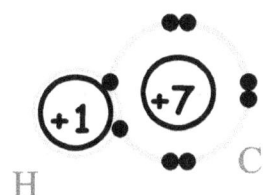

Explaining Chemical Processes

Activity 2.3: Reactions of Non-metal Oxides to form Acids

Do you Remember? **List 6 propositions of Arrhenius' Theory of Acids and Bases**

1. _____
2. _____
3. _____
4. _____
5. _____
6. _____

What's The Question? We found that non-metal oxides such as carbon dioxide and sulfur dioxide produced acid solutions in water.

How do water plus non-metal oxides react to form acids?

What Are We Thinking About?
- Pure water is a neutral substance. It is neither acid nor base.
- Non-metals have greater electronegativity than hydrogen. That is, they have a stronger ability to attract and hold electrons than hydrogen does. Non-metals and their oxides tend to attract electrons *away* from water, leaving H^+ ions.
- Each non metal in the list below undergoes three reactions:
  - i) Non-metals react with oxygen to form non-metal oxides;
  - ii) Non-metal oxides react with water to form acids;
  - iii) Acids dissolve in water, dissociating into hydrogen ions and polyatomic ions.

What Are We Doing?
1. Balance all of the equations in the list that follows.

2. Fill in the blanks with the names of the substances in each equation.

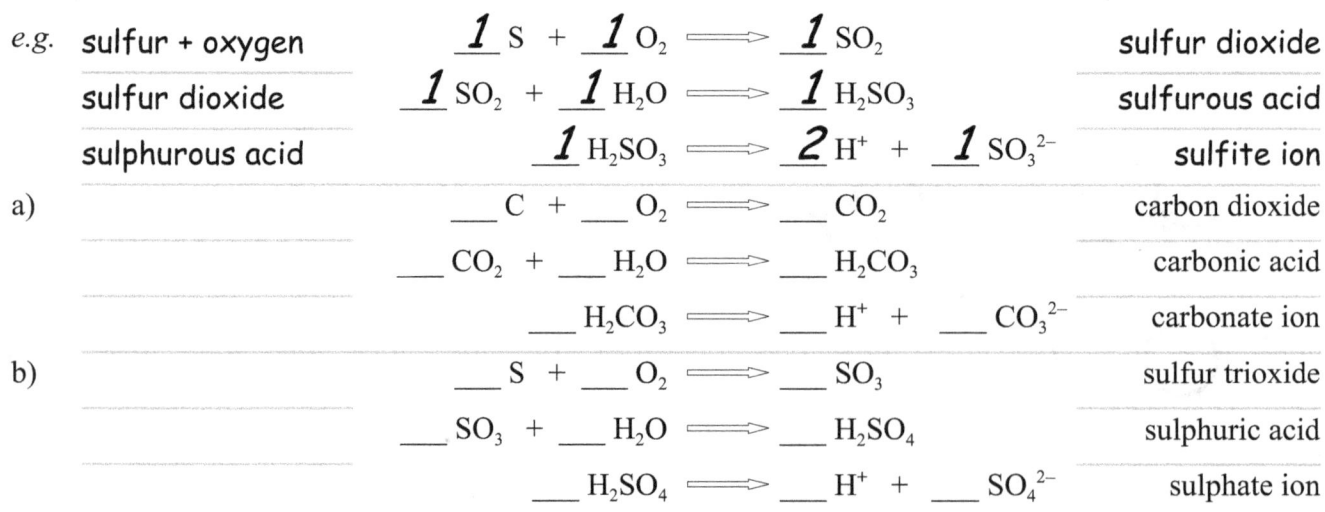

e.g. sulfur + oxygen	$\underline{1}\ S\ +\ \underline{1}\ O_2 \Longrightarrow \underline{1}\ SO_2$	sulfur dioxide
sulfur dioxide	$\underline{1}\ SO_2\ +\ \underline{1}\ H_2O \Longrightarrow \underline{1}\ H_2SO_3$	sulfurous acid
sulphurous acid	$\underline{1}\ H_2SO_3 \Longrightarrow \underline{2}\ H^+\ +\ \underline{1}\ SO_3^{2-}$	sulfite ion
a) _____	$\underline{\ \ }\ C\ +\ \underline{\ \ }\ O_2 \Longrightarrow \underline{\ \ }\ CO_2$	carbon dioxide
_____	$\underline{\ \ }\ CO_2\ +\ \underline{\ \ }\ H_2O \Longrightarrow \underline{\ \ }\ H_2CO_3$	carbonic acid
_____	$\underline{\ \ }\ H_2CO_3 \Longrightarrow \underline{\ \ }\ H^+\ +\ \underline{\ \ }\ CO_3^{2-}$	carbonate ion
b) _____	$\underline{\ \ }\ S\ +\ \underline{\ \ }\ O_2 \Longrightarrow \underline{\ \ }\ SO_3$	sulfur trioxide
_____	$\underline{\ \ }\ SO_3\ +\ \underline{\ \ }\ H_2O \Longrightarrow \underline{\ \ }\ H_2SO_4$	sulphuric acid
_____	$\underline{\ \ }\ H_2SO_4 \Longrightarrow \underline{\ \ }\ H^+\ +\ \underline{\ \ }\ SO_4^{2-}$	sulphate ion

c)

___ Br_2 + ___ $O_2 \Longrightarrow$ ___ Br_2O_5 — dibromine pentoxide

___ Br_2O_5 + ___ $H_2O \Longrightarrow$ ___ $HBrO_3$ — bromic acid

___ $HBrO_3 \Longrightarrow$ ___ H^+ + ___ BrO_3^- — bromate ion

d)

___ Cl_2 + ___ $O_2 \Longrightarrow$ ___ Cl_2O_3 — dichlorine trioxide

___ Cl_2O_3 + ___ $H_2O \Longrightarrow$ ___ $HClO_2$ — chlorous acid

___ $HClO_2 \Longrightarrow$ ___ H^+ + ___ ClO_2^- — chlorite ion

e)

___ Cl_2 + ___ $O_2 \Longrightarrow$ ___ Cl_2O_5 — dichlorine pentoxide

___ Cl_2O_5 + ___ $H_2O \Longrightarrow$ ___ $HClO_3$ — chloric acid

___ $HClO_3 \Longrightarrow$ ___ H^+ + ___ ClO_3^- — chlorate ion

f)

___ N_2 + ___ $O_2 \Longrightarrow$ ___ N_2O_5 — dinitrogen pentoxide

___ N_2O_5 + ___ $H_2O \Longrightarrow$ ___ HNO_3

nitric acid ___ $HNO_3 \Longrightarrow$ ___ H^+ + ___ NO_3^- — nitrate Ion

g)

___ N_2 + ___ $O_2 \Longrightarrow$ ___ N_2O_3

dinitrogen trioxide ___ N_2O_3 + ___ $H_2O \Longrightarrow$ ___ HNO_2 — nitrous acid

___ $HNO_2 \Longrightarrow$ ___ H^+ + ___ NO_2^- — nitrite ion

h)

___ P + ___ $O_2 \Longrightarrow$ ___ P_2O_5 — phosphorus pentoxide

___ P_2O_5 + ___ $H_2O \Longrightarrow$ ___ H_3PO_4

phosphoric acid ___ $H_3PO_4 \Longrightarrow$ ___ H^+ + ___ PO_4^{3-} — phosphate Ion

i)

___ H_2 + ___ $Cl_2 \Longrightarrow$ ___ HCl — hydrogen chloride

hydrochloric acid ___ $HCl \Longrightarrow$ ___ H^+ + ___ Cl^- — chloride ion

j)

___ H_2 + ___ $Br_2 \Longrightarrow$ ___ HBr — hydrogen bromide

hydrobromic acid ___ $HBr \Longrightarrow$ ___ H^+ + ___ Cl^- — bromide ion

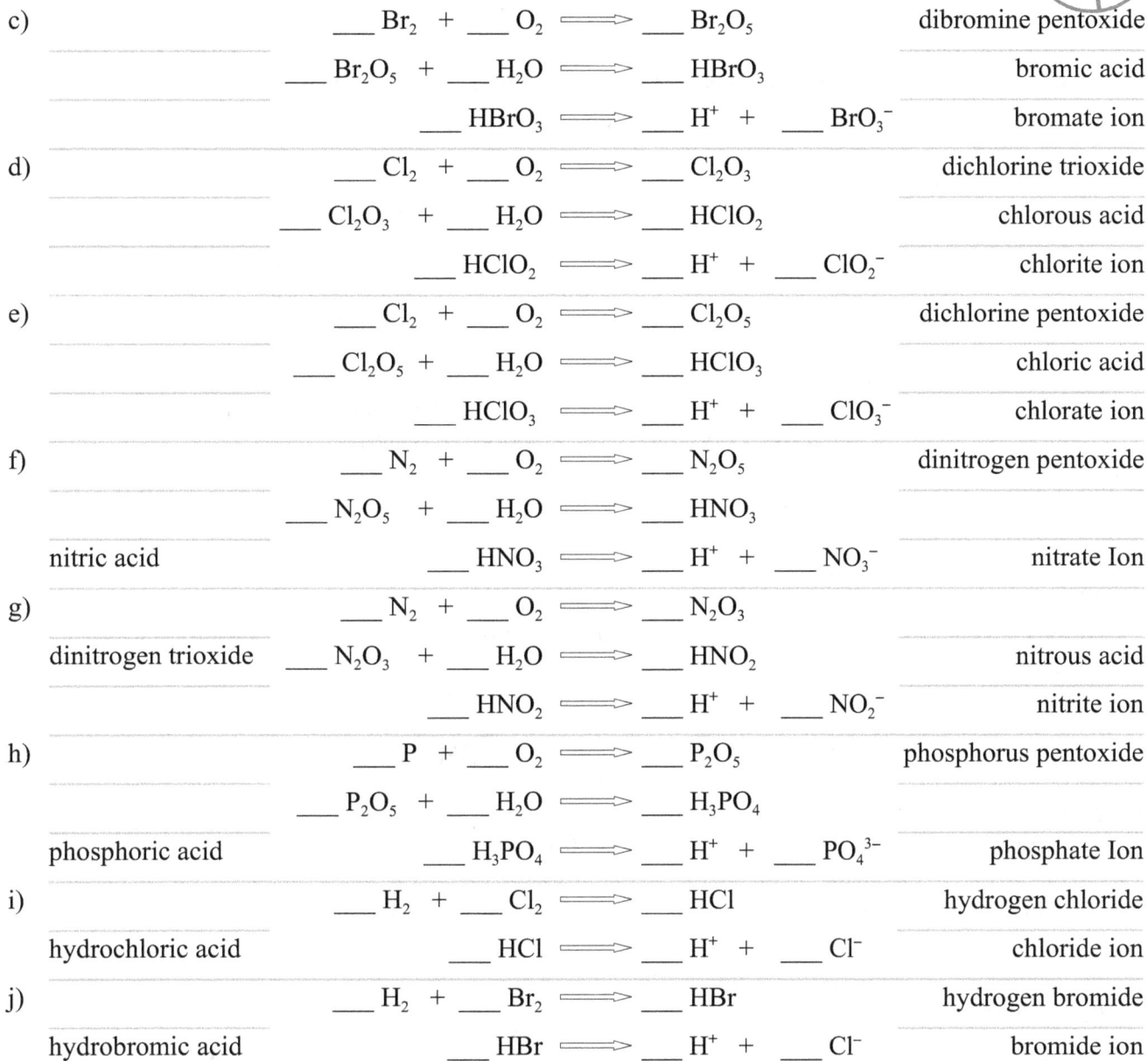

Questions For Later...
1. Which acids have no oxygen at all? How are their names different from the oxy acids?

2. Some acids have similar names, e.g. *nitric* and *nitrous*. Which acids appear to have more oxygen in them: the "*ic*" acids or the "*ous*" acids?

3. Which acids ("*ic*" or "*ous*") make "*ate*" ions? Which acids make "*ite*" ions?

Explaining Chemical Processes

Activity 2.4: Metal Oxides and Bases

Do you Remember? **List 6 propositions of Arrhenius' theory of Acids and Bases**

1. _____
2. _____
3. _____
4. _____
5. _____
6. _____

What's The Question? Group 1, 2, and 3 metals such as sodium and magnesium react with oxygen to form oxides. Solutions of metal oxides produce basic solutions in water.

How do water plus metal oxides react to form acids?

What Are We Thinking About?
- Pure water is a neutral substance. It is neither acid nor base.
- Metals have less electronegativity than oxygen. Therefore, metals *lose* electrons to oxygen, making oxide ions.
- When oxide ions meet water molecules, they combine to make two hydroxide ions.

magnesium $\underline{1}\ O^{2-} + \underline{1}\ H_2O \Longrightarrow \underline{2}\ OH^-$ hydroxide ions

What Are We Doing?
1. Balance *all* of the equations in the list that follows.

2. Fill in the blanks with the names of all substances except water and oxygen.

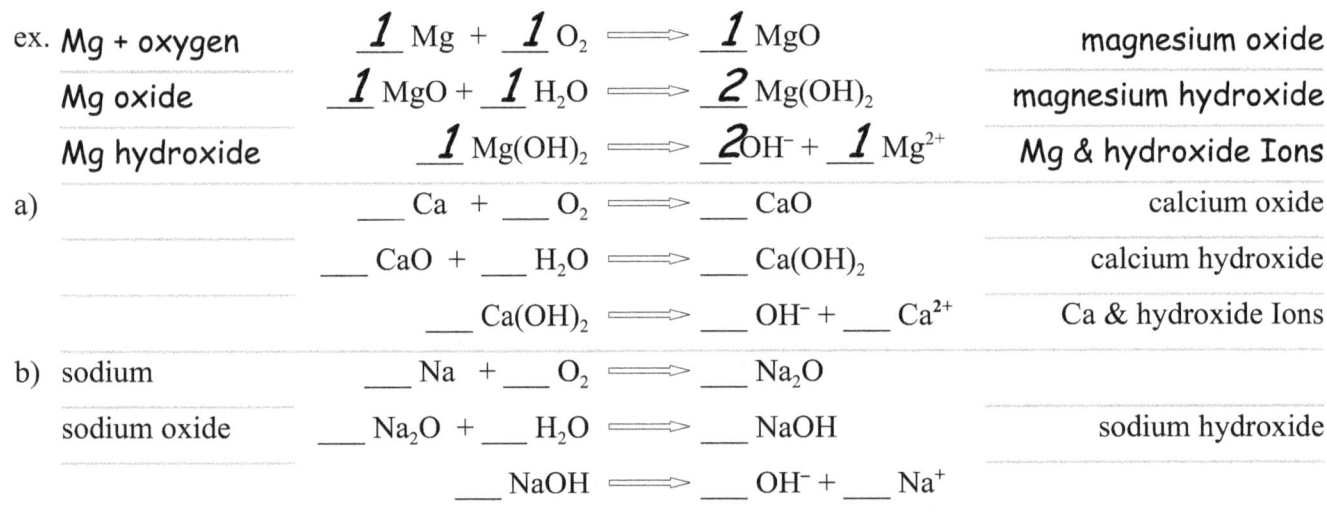

ex. Mg + oxygen	$\underline{1}\ Mg + \underline{1}\ O_2 \Longrightarrow \underline{1}\ MgO$		magnesium oxide
Mg oxide	$\underline{1}\ MgO + \underline{1}\ H_2O \Longrightarrow \underline{2}\ Mg(OH)_2$		magnesium hydroxide
Mg hydroxide	$\underline{1}\ Mg(OH)_2 \Longrightarrow \underline{2}OH^- + \underline{1}\ Mg^{2+}$		Mg & hydroxide Ions
a)	___ Ca + ___ O$_2$ $\Longrightarrow$ ___ CaO		calcium oxide
	___ CaO + ___ H$_2$O $\Longrightarrow$ ___ Ca(OH)$_2$		calcium hydroxide
	___ Ca(OH)$_2$ $\Longrightarrow$ ___ OH$^-$ + ___ Ca^{2+}		Ca & hydroxide Ions
b) sodium	___ Na + ___ O$_2$ $\Longrightarrow$ ___ Na$_2$O		
sodium oxide	___ Na$_2$O + ___ H$_2$O $\Longrightarrow$ ___ NaOH		sodium hydroxide
	___ NaOH $\Longrightarrow$ ___ OH$^-$ + ___ Na$^+$		

Metals, Non-Metals and pH

Name:
Date:

c)   ___ Li + ___ O_2 $\Longrightarrow$ ___ Li_2O lithium oxide

     ___ Li_2O + ___ H_2O $\Longrightarrow$ ___ LiOH

     ___ LiOH $\Longrightarrow$ ___ OH^- + ___ Li^+

d)   ___ Zn + ___ O_2 $\Longrightarrow$ ___ ZnO   zinc oxide

     ___ ZnO + ___ H_2O $\Longrightarrow$ ___ $Zn(OH)_2$ zinc hydroxide

     ___ $Zn(OH)_2$ $\Longrightarrow$ ___ OH^- + ___ Zn^{2+}

e)   ___ Ba + ___ O_2 $\Longrightarrow$ ___ BaO   barium oxide

     ___ BaO + ___ H_2O $\Longrightarrow$ ___ $Ba(OH)_2$ barium hydroxide

     ___ $Ba(OH)_2$ $\Longrightarrow$ ___ OH^- + ___ Ba^{2+}

f)   ___ Al + ___ O_2 $\Longrightarrow$ ___ Al_2O_3 aluminum oxide

     ___ Al_2O_3 + ___ H_2O $\Longrightarrow$ ___ $Al(OH)_3$

     ___ $Al(OH)_3$ $\Longrightarrow$ ___ OH^- + ___ Al^{3+}

g)   ___ K + ___ O_2 $\Longrightarrow$ ___ K_2O

     ___ K_2O + ___ H_2O $\Longrightarrow$ ___ KOH

     ___ KOH $\Longrightarrow$ ___ OH^- + ___ K^+

h)   ___ Ga + ___ O_2 $\Longrightarrow$ ___ Ga_2O_3

     ___ Ga_2O_3 + ___ H_2O $\Longrightarrow$ ___ $Ga(OH)_3$

     ___ $Ga(OH)_3$ $\Longrightarrow$ ___ OH^- + ___ Ga^{3+}

i)   ___ Na + ___ H_2O $\Longrightarrow$ ___ H_2 + ___ NaOH   sodium hydroxide

     ___ NaOH $\Longrightarrow$ ___ OH^- + ___ Na^+

j)   ___ Ca + ___ H_2O $\Longrightarrow$ ___ H_2 + ___ $Ca(OH)_2$ calcium hydroxide

     ___ $Ca(OH)_2$ $\Longrightarrow$ ___ OH^- + ___ Ca^{2+}

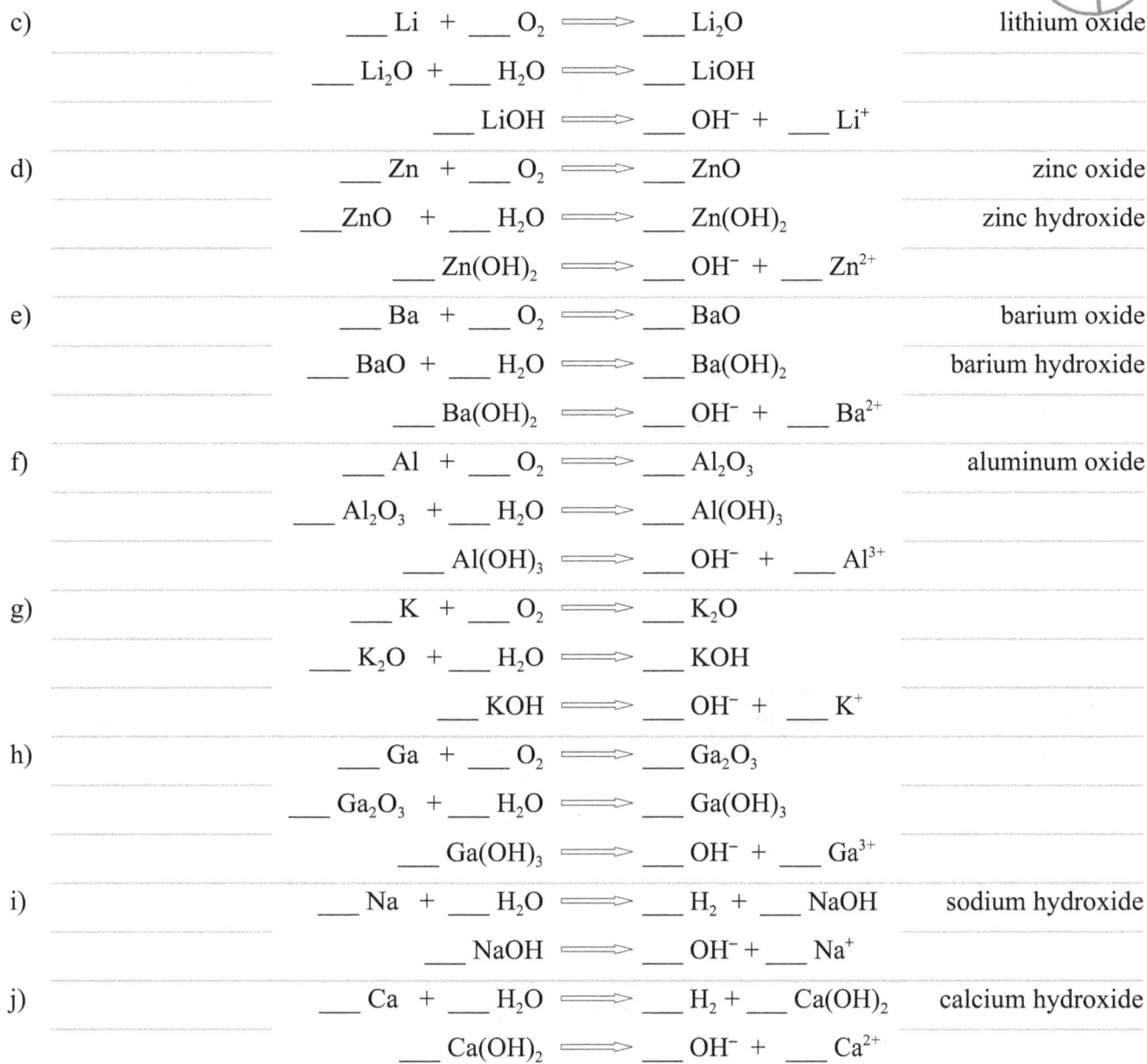

Questions For Later...

1.   Which metals reacted directly with water to form hydrogen gas plus a base?

2.   Which metals produced the greatest amount of hydroxide?

Explaining Chemical Processes

Lab 2.5: Testing the Strength of Acids and Bases

Do you Remember? **Write the 6 propositions of Arrhenius' Theory of Acids and Bases**

1. _____
2. _____
3. _____
4. _____
5. _____
6. _____

What's The Question? The hydrogen ion, H^+, is what makes acid acidic. *How many drops of strong* pH 2 acid *must we mix with neutral water to make* pH 3 acid? *How many drops to make pH 4 acid? pH 5 acid? pH 6 acid?*

What Are We Doing?

1. Obtain 500 mL of distilled water. Add 5 mL of Universal Indicator. The solution should be neutral (pH 7.0, yellow-green)
2. Measure out 100 mL of the neutral solution into 5 clean dry beakers.. Label the beakers pH 3, pH 4, pH 5, pH 6 and pH 7 (see diagram opposite page).
3. Obtain 100 mL of strong acid, pH 2.0 from your teacher.
Predict how many drops of the strong pH 2 acid you must add to each beaker to make the pH indicated. **Explain** your prediction.
4. One drop at a time, counting each drop, add the strong pH 2 acid to the neutral water to make pH 3. *Stir after adding each drop.*
5. Repeat the procedure to make solutions pH 4, pH 5 and pH 6.
Observe the number of drops and record this number on the accompanying diagram. **Explain** your results.

What Are We Thinking About?

Caution: acids are corrosive *Never take your goggles off during this lab: even if you finish, others may still be working.*

*   **pH** is a scale used to measure the relative strength of hydrogen. In the original German, pH meant *"potency of hydrogen"*

    Pure neutral water is **pH = 7**
    Strong bases are **pH = 14**
    Strong acids are **pH = 0**

*   Your teacher will provide a series of acid and base samples to show you the colour of Universal Indicator at each pH.

Questions For Later...

1. If a sample of solution Y pH 3 has 10 "acid particles", how many "acid particles" would solution X have? Colour them black.

2. If Y has 10 particles, how many "acid particles" would solution Z have? Colour them black.

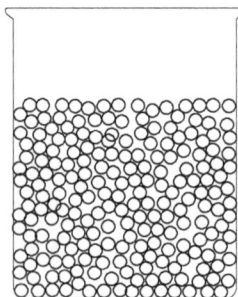
X   pH 2

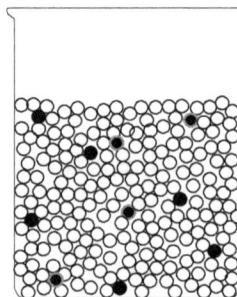
Y   pH 3

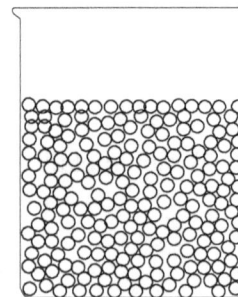
Z   pH 4

Metals, Non-Metals and pH

Name:

Date:

Focus Question: Write the question that you are trying to answer.

1   **Predict** how many drops of strong (pH 2 acid) you must add to 100 mL neutral water ...

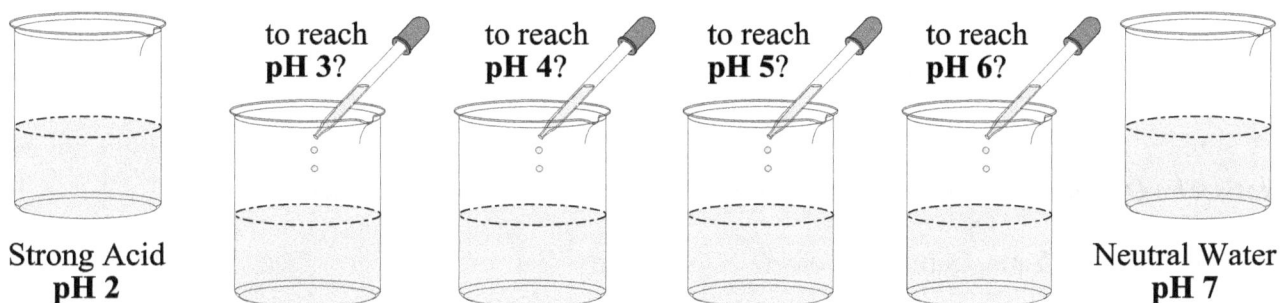

Strong Acid
pH 2

to reach **pH 3**?

to reach **pH 4**?

to reach **pH 5**?

to reach **pH 6**?

Neutral Water
pH 7

2   **Explain** your prediction, using sentences and pictures.

3   **Observe** and make records of the changes in pH of the acids that you diluted.

4   **Explain** the relationship between the pH scale and the dilution of the acid.

Explaining Chemical Processes

Lab 2.6: Testing the pH of Common Substances

Do you Remember? **List the 6 propositions of Arrhenius' theory of Acids and Bases**

1. _____
2. _____
3. _____
4. _____
5. _____
6. _____

What's The Question? You have heard a great deal about acids, bases, and pH.

What is the pH of substances in your house?

What Are We Doing?

1. Make a list of 20 substances found in your house. **Predict** the pH of each substance.

2. **Explain** your prediction, e.g.

   "I think that milk has 1/10 as much H^+ as vinegar, so it will have a pH of ____."

3. **Observe** the pH by testing with pH paper. Use about 5 mm of test paper for each test.

4. **Explain** your measurements, e.g..

   "Milk has a pH of ____, vinegar's pH is ____. Therefore milk has ____ × more H^+ than vinegar.

What Are We Thinking About?

- Vinegar is a common household acid. All other acids can be compared to vinegar. Describe each acid as having 10, 100 or 1000 times more H^+, or as having 1/10, 1/100, or 1/1000 as much H^+ as vinegar.

- A paste of baking soda (sodium bicarbonate) in water can be used as a standard base, at pH 9. All other bases can be compared to this solution. Describe every other base as having 10, 100 or 1000 times more OH^-, or as having 1/10, 1/100, or 1/1000 as much OH^-.

Questions For Later...

1. What is the pH of the strongest acid in your house? What safety precautions were on its package?

2. What is the pH of the strongest base in your house? What safety precautions were on its package?

3. What is the strongest acid and strongest base that you regularly put in contact with your body?

Metals, Non-Metals and pH

Name:

Date:

	Substance	Predict pH	Explain: "I think there is 10× more H⁺ or OH⁻"	Observe pH	Explain: "there is 100× more H⁺ than there is in vinegar
1	Vinegar				
2	Laundry detergent				
3					
4					
5					
6					
7					
8					
9					
10					
11					
12					
13					
14					
15					
16					
17					
18					
19					
20					

Quiz 2.7: Metals, Non-metals and pH

1 In the short form of the periodic table below, the transition metals have been left out.

Mark each square on the table as metal (**M**), non-metal (**N**) or metalloid(**O**).

Date: _____ / 5

2 Each of the elements shown below are reacted with oxygen, and dissolved in water.

Li				C	N		
Na	Mg	Al		P	S	Cl	
K	Ca					Br	

Mark each of those elements as making acidic solutions (A), or basic solutions (B).

Date: _____ / 5

3 When magnesium metal burns in oxygen, the compound magnesium oxide MgO is synthesized. If the MgO is dissolved in water, it reacts to form the base magnesium hydroxide, $Mg(OH)_2$
Draw molecular diagrams and write a balanced chemical equation for each reaction.

Date: _____ / 5

4 If sulfur burns in oxygen, it produces a compound sulfur dioxide, SO_2. When the sulfur dioxide is further dissolved in water, sulfurous acid H_2SO_3 is produced. Draw molecular diagrams and write a balanced chemical equation for each of the reactions.

Date: _____ / 5

The Grade Ten Daily

Quiz 2.7: Metals, Non-metals and pH   Name:

---

5   The substances in the list below are either acids (A) or bases (B). Which ones are which?

a) _____ H_2SO_4 f) _____ HNO_3

b) _____ $Mg(OH)_2$ g) _____ NaOH

c) _____ H_2CO_3 h) _____ $Al(OH)_3$

d) _____ KOH   i) _____ $HClO_3$

e) _____ $Ca(OH)_2$ j) _____ H_2O

Explain why you made your choices.

Date: _____   / 5

---

6   Examine each compound in the list below. Suppose that each one was dissolved in water.

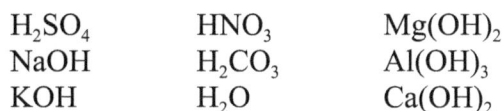

H_2SO_4	HNO_3	$Mg(OH)_2$
NaOH	H_2CO_3	$Al(OH)_3$
KOH	H_2O	$Ca(OH)_2$

Which ones could make a solution

pH 4   _____

pH 7   _____

pH 10 _____

Date: _____   / 5

---

7   When dissolved in water, each of these compounds will dissociate into ions. Write the equation for the dissociation equation.

e.g. $HCl \Longrightarrow H^+ + Cl^-$

a) $H_2SO_4 \Longrightarrow$

b) $HNO_3 \Longrightarrow$

c) $Mg(OH)_2 \Longrightarrow$

d)   NaOH $\Longrightarrow$

c)   HBr $\Longrightarrow$

Date: _____   / 5

---

8   Name each of the following oxides. When further reacted with water, these could form either an acid (A) or a base (B). Mark each compound A or B.

	Chemical name	*A or B*
a) MgO	_____	_____
b) SO_2	_____	_____
c) NO_2	_____	_____
d) K_2O	_____	_____
e) Al_2O_3	_____	_____

Explain your answers.

Date: _____   / 5

Quiz 2.7: Metals, Non-metals and pH   **Name:**

9   A small sample of acid Y, pH 6, has 6 "acid particles" in it. How many "acid particles has solution X, pH 5?

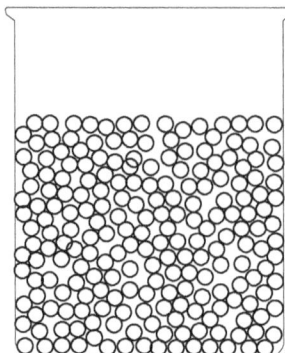

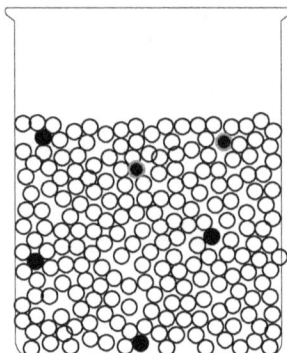

X   pH 5   **Y   pH 6**

Explain your answer.

Date: / 5

10   A tiny sample of base Q, pH 9, has 1 "base particle" in it. How many "base particles" would be found in solution R, pH 10?

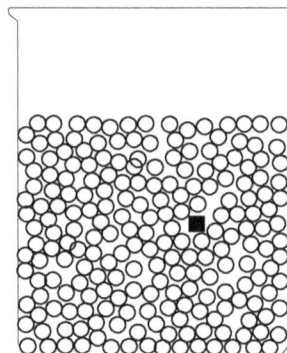

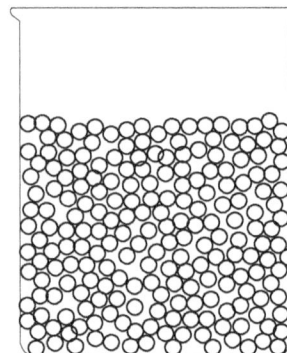

Q   pH 9   **R   pH 10**

Explain your answer.

Date: / 5

11   Write the name of each substance in the list below on the pH scale shown.

Neutral
Weak acid
Strong acid
Weak base
Strong base

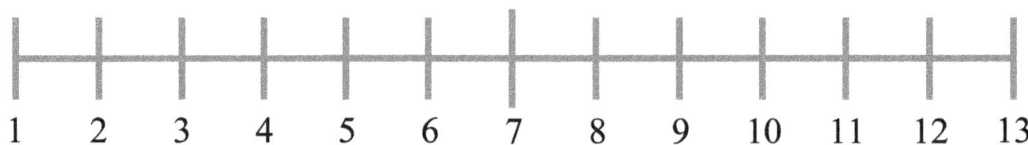

Date: / 5

12   Write the name of each substance in its correct place on the pH scale shown.

Lemonade
Water
Vinegar
Hand soap
Laundry detergent

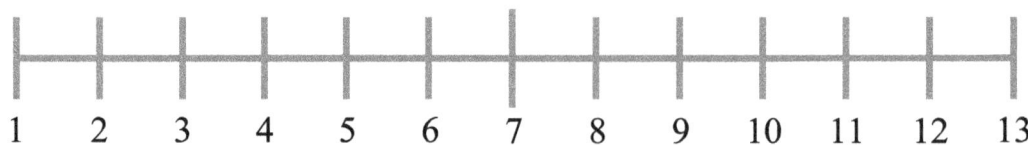

Date: / 5

13 A tiny sample of acid Y, pH 5, has 40 "acid particles" in it. Sample X has only 4 "acid particles". What is the pH of X ?

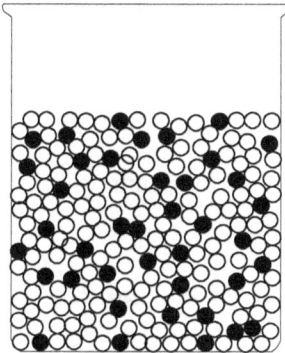

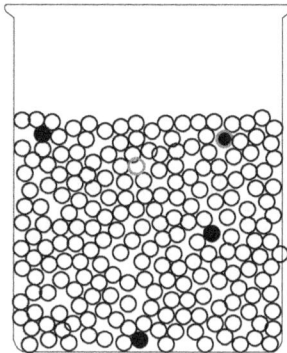

X   pH 5   **Y   pH = ?**

Explain your answer.

Date: / 5

14 A tiny sample of base R, pH 11, contains 30 "base particles". Sample Q has only 3 "base particles". What is the pH of Q?

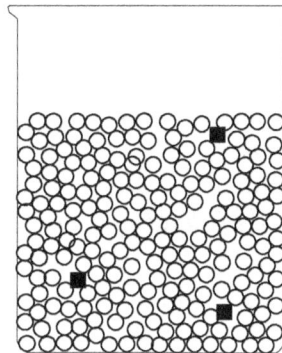

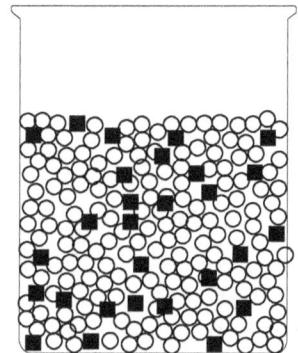

Q   pH = ?   **R   pH 11**

Explain your answer.

Date: / 5

15 Write the name of each substance in the list below on the pH scale shown.

Neutral
Weak acid
Strong acid
Weak base
Strong base

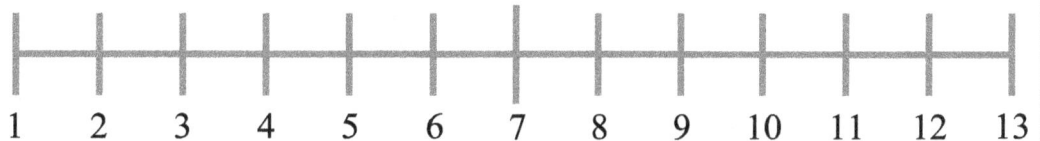

Date: / 5

16 Write the name of each substance in its correct place on the pH scale shown.

Shampoo
Drain cleaner
Lemon juice
Ketchup
Water

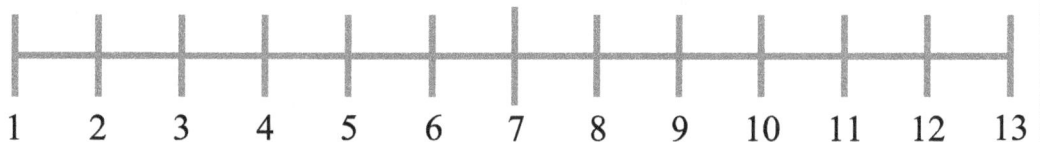

Date: / 5

Explaining Chemical Processes

Lab 3.1: Acids, Bases, and Neutralization

Do you Remember? Write the 6 propositions of Arrhenius Theory of Acids and Bases.

1. _____
2. _____
3. _____
4. _____
5. _____
6. _____

What's The Question? Hydrochloric Acid and Sodium Hydroxide are the most common acid and base in the chemistry lab. *What products are made when an acid and base react together?*

What Are We Doing? Before the lab, obtain
    25.0 mL of 0.10 M HCl
    25.0 mL of 0.10 M NaOH
    Purple Cabbage Extract indicator (PCE)

1. Add 5 mL of PCE to each solution. Note the colours.

2. Pour 20.0 mL of HCl into a 100 mL beaker. **Predict** the volume of NaOH needed to turn the solution neutral (navy blue). **Explain** your prediction.

3. Use a dropper to add NaOH to the HCl solution in the beaker. Add small quantities at a time, and swirl. If you add too much NaOH, add a drop of HCl solution.

4. When you have reached pH 7 (navy blue), taste a drop or two of the solution.

5. Write **Observations** and **Explain** your results.

What Are We Thinking About?

1. 1 mole is a number used by chemists to count molecules. 1 mole of any substance has the same number of molecules

2. 1 M NaOH has *1 mole of NaOH* dissolved in one litre of solution. 1 M HCl contains *1 mole of HCl* dissolved in one litre of solution.

3. Equal volumes of the solutions above contain equal numbers of molecules of HCl and NaOH

4. Purple cabbage juice is:
    purple   @ pH 6
    navy blue   @ pH 7
    teal blue   @ pH 8.

Questions For Later...

1. What substances appear to be produced by the reaction of an acid and a base?

2. What products might be produced if KOH (Potassium Hydroxide) was used to neutralize HCl (Hydrochloric Acid)?

3. What combination of acid and base might produce Lithium Bromide?

Reactions of Acids and Bases

Name:

Date:

Focus Question: Write the question that you are trying to answer.

Before the experiment...
Complete this diagram to show
four different representations of
the chemical change you predict.

In your pictures, chemical
formulae, and masses, you must
show that no matter has been
created or destroyed.

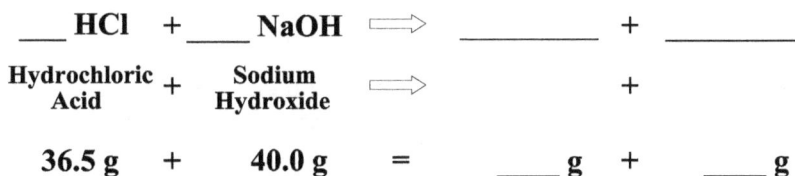

H Cl   Na O H ⟹

___ HCl   + ___ NaOH ⟹ _____ + _____

Hydrochloric + Sodium ⟹ +
Acid   Hydroxide

36.5 g + 40.0 g = ___ g + ___ g

---

1   **Predict:** What volume of NaOH do you expect
to use to exactly neutralize 20.0 mL of HCl?
Predict: What products will form?

2   **Explain** your prediction, using both
particle diagrams and sentences.

3   **Observe** the volume of NaOH added to
neutralize the HCl. **Observe** the taste of the
product from the reaction.

4   **Explain** your observations, using both
diagrams and sentences.

Explaining Chemical Processes

Lab 3.2: More Neutralization Reactions

Do you Remember? Write 6 propositions of Arrhenius' Theory of Acids and Bases

1. _____
2. _____
3. _____
4. _____
5. _____
6. _____

What's The Question? Carbonic acid and calcium hydroxide are able to neutralize each other.
What products are formed in this reaction?

What Are We Doing?

1. **Predict** the products of the two reactions. **Explain** your predictions, using both balanced chemical equations and particle diagrams.

2. Obtain 10 mL of $Ca(OH)_2$ solution, and add 5 drops of universal indicator.

3. Using a straw, bubble your breath through the solution until you achieve a neutral green colour. Continue bubbling until the solution reaches pH 5.

4. **Observe** and **Explain** the results of both procedures

Your teacher may conduct a second demonstration, using calcium hydroxide and sulfuric acid.

What Are We Thinking About?

1. What is the pH of neutral water?

2. What colour will the universal indicator be when there are equal amounts of acid and base, that is, neutral water?

3. What acid is made when Carbon Dioxide dissolves in water?

Questions For Later...

1. Identify the chemical name of the two salts produced in these two reactions

2. Describe the limewater test for carbon dioxide. If you can't recall using it yourself, look it up in another text book

3. What two chemical reactions must occur during the limewater test for carbon dioxide?

Reactions of Acids and Bases

Name:

Date:

Focus Question: Write the question that you are trying to answer.

Before the experiment...
Complete this diagram to show four different representations of the chemical change you predict.

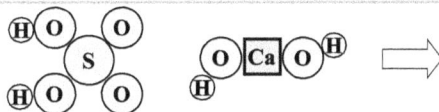

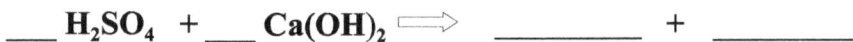

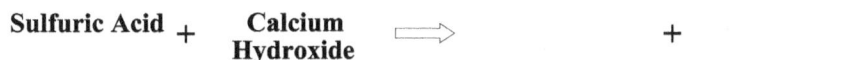

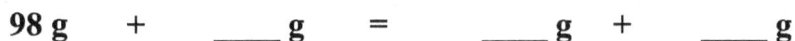

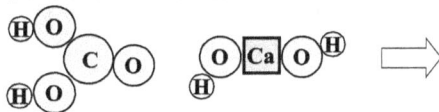

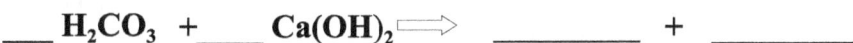

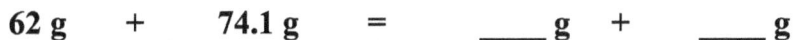

___ H_2SO_4 + ___ $Ca(OH)_2$ ⟹ _____ + _____

Sulfuric Acid + Calcium Hydroxide ⟹ +

98 g + ____ g = ____ g + ____ g

1 Predict and Explain	2 Observe and Explain

Before the experiment...
Complete this diagram to show four different representations of the chemical change you predict.

___ H_2CO_3 + ___ $Ca(OH)_2$ ⟹ _____ + _____

Carbonic Acid + Calcium Hydroxide ⟹ +

62 g + **74.1 g** = ____ g + ____ g

1 Predict and Explain	2 Observe and Explain

Explaining Chemical Processes

Lab 3.3: Reactions of Acids and Metals

Do you Remember? **Write 6 propositions of Arrhenius' Theory of Acids and Bases**

1. _____
2. _____
3. _____
4. _____
5. _____
6. _____

What's The Question? When a metal reacts with an acid, one of the products is a gas. The other product cannot be seen until the water and the remaining acid have been driven off.
What mass of Aluminum Chloride is produced when 0.54 g of Al reacts with HCl?

Your teacher may do this experiment

1. **Predict** the mass of $AlCl_3$ produced from 0.54 g of Al. **Explain** your prediction, using diagrams and sentences.

2. Obtain a clean, dry 250 mL flask, and write your name on the white spot with pencil. Weigh, and record the mass.

3. *Tare* the empty flask (zero the balance with the flask on it). Add bits of Al foil until the mass of Al is very nearly 0.54 g .

4. Add 100 mL of 1.0 M HCl.

5. Place the flask on a wire gauze, on a hot plate. Warm the HCl gently until the reaction begins. *Turn off the hot plate.* Collect some of the gas in an inverted test tube, and bring it to a Bunsen burner as shown.

6. When the reaction is over, place the flask in an oven until the liquid has evaporated. Overnight is usually enough.

7. Measure the mass of the cool, dry flask and residue. Examine the residue. Write your **observations** and **explain** the results.

What Are We Thinking About?

Caution! Hydrochloric Acid is corrosive. Reaction generates considerable heat.

1. Acid + metal always reacts to produce H_2 gas plus a salt (metal + acid radical).

2. In this reaction, one element (the metal) takes the place of another element (hydrogen) in the compound. Because the metal *displaces* the hydrogen, this is a *single displacement reaction*

3 Salts prepared in this way usually contain extra water molecules in the crystals!!

Questions For Later...

1. What is the identity of the dry residue? Give two reasons why you believe that.

2. Would your measured mass of $AlCl_3$ be heavier or lighter than predicted, if the water was not completely driven off? Explain your thinking.

Reactions of Acids and Bases

Name:

Date:

Focus Question: Write the question that you are trying to answer.

Before the experiment...
Complete this diagram to show four different representations of the chemical change you predict.

You need to *balance* this equation before you calculate the masses!

___HCl   +   ___Al   ⟹   _____   +   _____

Hydrochloric + Aluminum ⟹ +
Acid

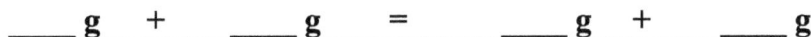

____ g   +   ____ g   =   ____ g   +   ____ g

---

1 *Predict*	2 *Explain*

---

3   *Observe*

Mass of empty beaker   _____
Mass of Aluminum   _____
Mass of beaker + dry residue   _____
Mass of dry residue   _____

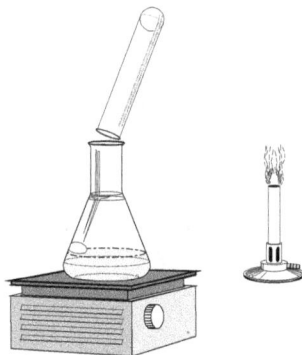

4   *Explain*

Explaining Chemical Processes

Activity 3.4: Balancing Neutralization Reactions
Provide both the word equation and the balanced chemical equation for each reaction.

1.	___ KOH + ___ HCl $\Longrightarrow$ ___ H_2O + ___ KCl
	potassium hydroxide + hydrochloric acid $\Longrightarrow$ water + potassium chloride
2.	___ $Ca(OH)_2$ + ___ HBr $\Longrightarrow$ ___ H_2O + ___ $CaBr_2$
	$\Longrightarrow$
3.	___ $Al(OH)_3$ + ___ HI $\Longrightarrow$ ___ H_2O + ___ AlI_3
	$\Longrightarrow$
4.	___ LiOH + ___ H_2SO_4 $\Longrightarrow$ ___ H_2O + ___ Li_2SO_4
	$\Longrightarrow$
5.	___ $Cu(OH)_2$ + ___ HNO_3 $\Longrightarrow$ ___ H_2O + ___ $Cu(NO_3)_2$
	$\Longrightarrow$
6.	___ H_3PO_4 + ___ $Mg(OH)_2$ $\Longrightarrow$ ___ H_2O + ___ $Mg_3(PO_4)_2$
	$\Longrightarrow$
7.	$\Longrightarrow$
	carbonic acid + sodium hydroxide $\Longrightarrow$ water + sodium sarbonate
8.	$\Longrightarrow$
	sulfurous acid + calcium hydroxide $\Longrightarrow$ water + calcium sulfite
9.	$\Longrightarrow$
	perchloric acid + iron (III) hydroxide $\Longrightarrow$ water + iron (III) perchlorate
10.	$\Longrightarrow$
	barium hydroxide + phosphoric acid $\Longrightarrow$ water + barium phosphate
11.	$\Longrightarrow$
	aluminum hydroxide + acetic acid $\Longrightarrow$ water + aluminum acetate
12.	$\Longrightarrow$
	ammonium hydroxide + nitric acid $\Longrightarrow$ water + ammonium nitrate

Reactions of Acids and Bases

Name:

Date:

Activity 3.5: Balancing Reactions of Metals and Acids

Provide both the word equation and the balanced chemical equation for each reaction.

1. ___ Zn + ___ $HCl \Longrightarrow$ ___ H_2 + ___ $ZnCl_2$

   zinc + hydrochloric acid $\Longrightarrow$ hydrogen gas + zinc chloride

2. ___ Mg + ___ $HBr \Longrightarrow$ ___ H_2 + ___ $MgBr_2$

   $\Longrightarrow$

3. ___ Al + ___ $HI \Longrightarrow$ ___ H_2 + ___ AlI_3

   $\Longrightarrow$

4. ___ Li + ___ $H_2SO_4 \Longrightarrow$ ___ H_2 + ___ Li_2SO_4

   $\Longrightarrow$

5. ___ Cu + ___ $HNO_3 \Longrightarrow$ ___ H_2 + ___ $Cu(NO_3)_2$

   $\Longrightarrow$

6. ___ H_3PO_4 + ___ $Mg \Longrightarrow$ ___ H_2 + ___ $Mg_3(PO_4)_2$

   $\Longrightarrow$

7. $\Longrightarrow$

   carbonic acid + sodium $\Longrightarrow$ hydrogen gas + sodium carbonate

8. $\Longrightarrow$

   sulfurous acid + calcium $\Longrightarrow$ hydrogen gas + calcium sulfite

9. $\Longrightarrow$

   chloric acid + iron $\Longrightarrow$ hydrogen gas + iron chlorate

10. $\Longrightarrow$

    barium + phosphoric acid $\Longrightarrow$ hydrogen gas + barium phosphate

11. $\Longrightarrow$

    aluminum + acetic acid $\Longrightarrow$ hydrogen gas + aluminum acetate

12. $\Longrightarrow$

    zinc + nitric acid $\Longrightarrow$ hydrogen gas + zinc nitrate

Lab 3.6: Rate of Reaction of Metal with Acid

Do you Remember? Write the 6 propositions of Arrhenius' theory of acids and bases.

1. _____
2. _____
3. _____
4. _____
5. _____
6. _____

What's The Question? In a previous lab, you observed the reaction of metal with acid. The products were Hydrogen gas plus a metal salt of the acid. *How will the pH change as the reaction proceeds? How does pH affect the speed of the reaction? What other factors might affect the rate of reaction?*

What Are We Thinking About?
Caution: Hydrochloric acid and calcium metal can injure eyes and skin. Avoid contact with skin. *Never take your goggles off during this lab: even if you finish, others may still be working.*

What Are We Doing?
1. You will start with acid having a pH of about 1. **Predict** how the pH will change, and how the speed of the reaction will change as the pH changes. **Explain** your prediction.
2. Assemble the apparatus shown at right. Your teacher will provide 20 mL of 0.10 M HCl, and a piece of Ca metal. Using forceps, place the Ca inside the flask, and replace the stopper. Start timing when the reaction begins.
3. **Observe** and record the pH and volume of H_2 gas collected every 30 s. **Explain** your results.

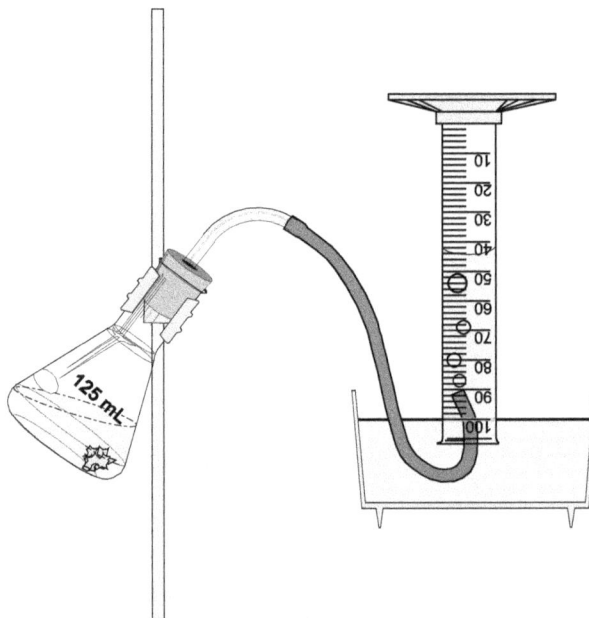

Questions For Later...
1. Does pH increase or decrease as the acid is used up in the reaction? Explain.

2. Which causes the faster reaction: pH 1, or pH 4? Explain

3. What would happen to the rate of the reaction if the test tube was kept in an ice water bath?

4. What would happen to the rate if you added 20 mL of water to the acid before you started?

Reactions of Acids and Bases

Focus Question: Write the question that you are trying to answer.

1 **Predict**

2 **Explain**

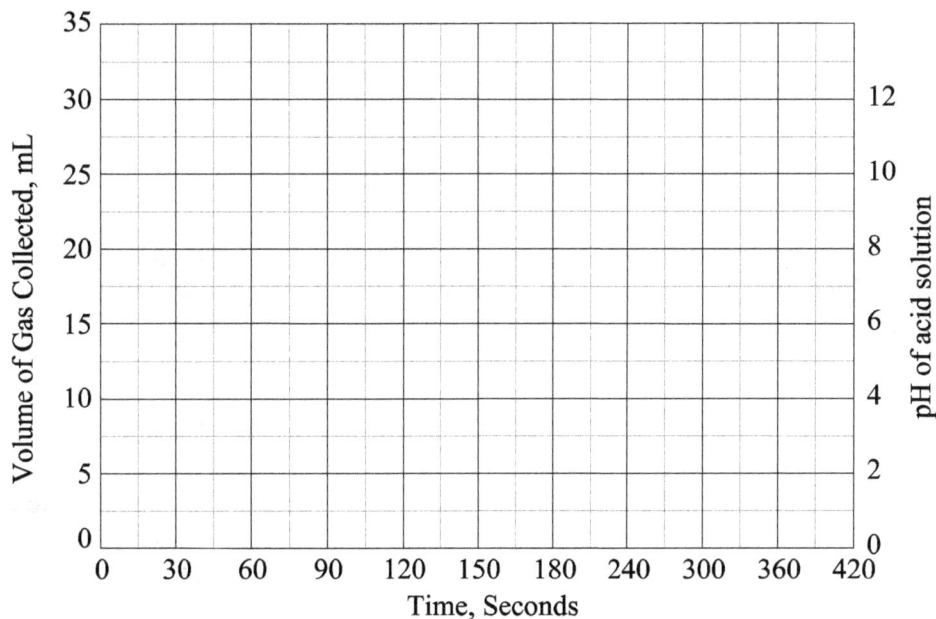

3 **Observe**

4 **Explain**

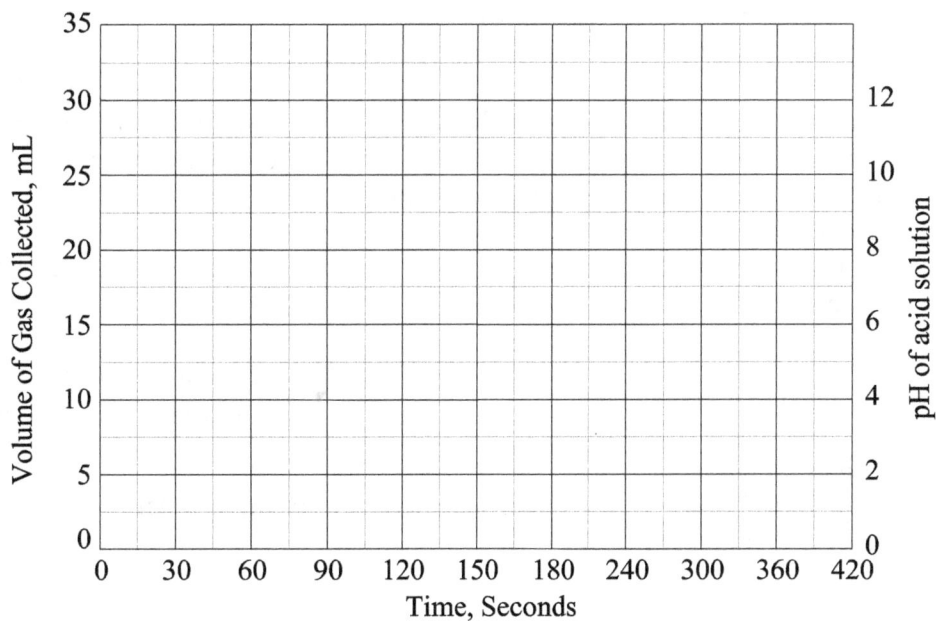

Explaining Chemical Processes

Lab 3.7: Reaction of Acids with Carbonates

Do you Remember? **Define 6 relevant concepts to this lab.**

1. _____
2. _____
3. _____
4. _____
5. _____
6. _____

What's The Question? When an acid reacts with a carbonate, three new products are formed. Two of the products are always the same. One product is different for each acid and carbonate.
What are the products formed when an acid and a carbonate react?

What Are We Doing?

You will be given 2 acids and 2 carbonates:
  sulfuric acid H_2SO_4
  hydrochloric acid   HCl.
  calcium carbonate $CaCO_3$
  sodium carbonate Na_2CO_3 .

In four separate experiments, you will combine each acid with each carbonate.

1. **Predict** the products formed with each combination. **Explain** your predictions, using diagrams, balanced equations, and sentences.
2. Do the experiment. Test the gas produce. Is it CO_2 ? **Observe** and **Explain** your observations.

What Are We Thinking About?

Caution: Hydrochloric and sulfuric acids can injure eyes, skin, and damage clothing. *Never take your goggles off during this lab: even if you finish, others may still be working.*

1. Carbonates contain metal ions plus carbonate ions.

2. Carbonic acid (H_2CO_3) is formed when carbon dioxide reacts with water. If carbonic acid decomposed, what products would you expect?

3. Limestone, chalk and marble are three forms of calcium carbonate. Laundry detergent is mostly sodium carbonate.

4. The test for CO_2: collect some of the gas produced, and shake it with limewater.

Questions For Later...
1. Vinegar is 5% acetic acid (hydrogen acetate). What products would you expect if you mixed vinegar and laundry detergent?

2. Egg shells contain calcium carbonate. If an egg was soaked in a glass of vinegar for a week, what do you predict would happen to the shell? Try it!.(replace with fresh vinegar each morning!!)

Reactions of Acids and Bases

Name:

Date:

Focus Question: Write the question that you are trying to answer.

Predict the products of the reaction, and use the completed representations of the chemical reaction to **explain** your prediction.

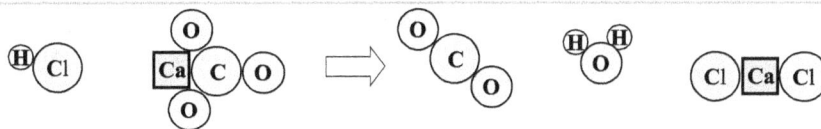

You need to *balance* this equation before you calculate the masses!

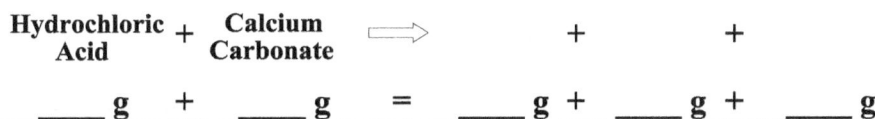

___ HCl   +___ CaCO₃ ⟹ _____ + _____ +_____

Hydrochloric Acid + Calcium Carbonate ⟹ _____ + _____

____ g   +   ____ g   =   ___ g + ____ g + ____ g

Predict and **explain** as above.

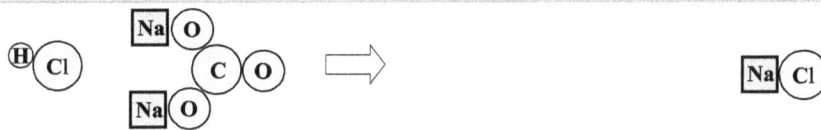

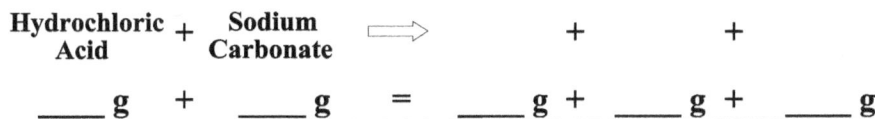

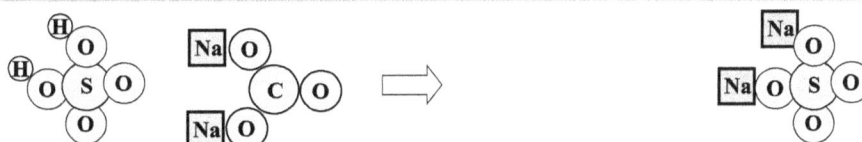

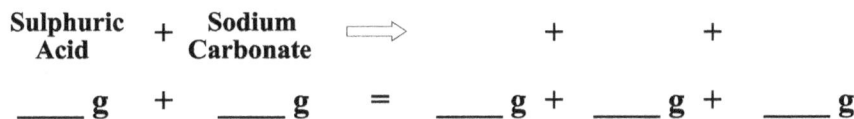

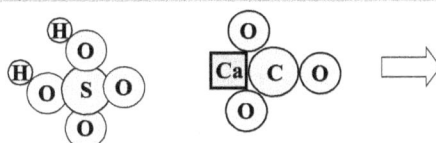

___ HCl   +___ Na₂CO₃ ⟹ _____ + _____ +_____

Hydrochloric Acid + Sodium Carbonate ⟹ _____ + _____

____ g   +   ____ g   =   ___ g + ____ g + ____ g

Predict and **explain** as above.

___ H₂SO₄   +___ Na₂CO₃ ⟹ _____ + _____ +_____

Sulphuric Acid + Sodium Carbonate ⟹ _____ + _____

____ g   +   ____ g   =   ___ g + ____ g + ____ g

Predict and **explain** as above.

___ H₂SO₄   +___ CaCO₃ ⟹ _____ + _____ +_____

Sulphuric Acid + Calcium Carbonate ⟹ _____ + _____

____ g   +   ____ g   =   ___ g + ____ g + ____ g

© Ross Lattner Publishing   77   www.rosslattner.com

Quiz 3.8: Reactions of Acids and Bases

1 According to Arrhenius' theory, when an acid dissolves it dissociates, producing H^+ ion. Complete all four representations of the acid dissociation reaction.

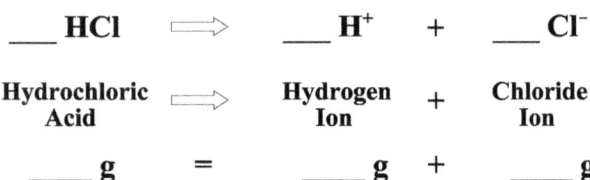

(H)(Cl) $\Rightarrow$

___ HCl $\Rightarrow$ ___ H^+ + ___ Cl^-

Hydrochloric Acid $\Rightarrow$ **Hydrogen Ion** + **Chloride Ion**

___ g = ___ g + ___ g

Date: ___ / 5

2 According to Arrhenius' theory, when an base dissolves it dissociates, producing OH^- ion. Complete all four representations of the base dissociation reaction.

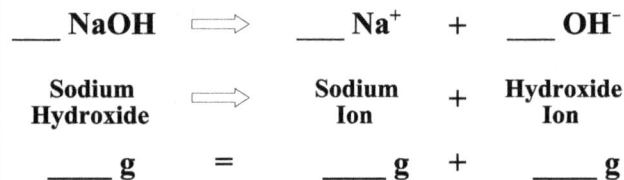

(Na)(O)(H) $\Rightarrow$

___ NaOH $\Rightarrow$ ___ Na^+ + ___ OH^-

Sodium Hydroxide $\Rightarrow$ **Sodium Ion** + **Hydroxide Ion**

___ g = ___ g + ___ g

Date: ___ / 5

3 According to Arrhenius, acid and base react to form water plus a salt. Complete all four representations of this *neutralization* reaction.

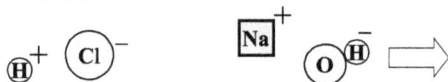

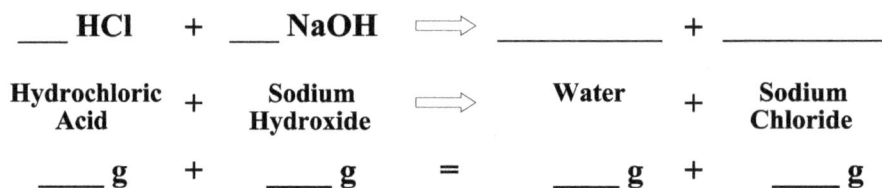

$(H)^+$ $(Cl)^-$ $[Na]^+$ $(O)(H)^-$ $\Rightarrow$

___ HCl + ___ NaOH $\Rightarrow$ _____ + _____

Hydrochloric Acid + **Sodium Hydroxide** $\Rightarrow$ **Water** + **Sodium Chloride**

___ g + ___ g = ___ g + ___ g

Date: ___ / 5

4 Complete all four representations of this *neutralization* reaction.

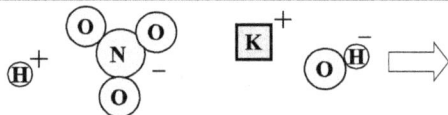

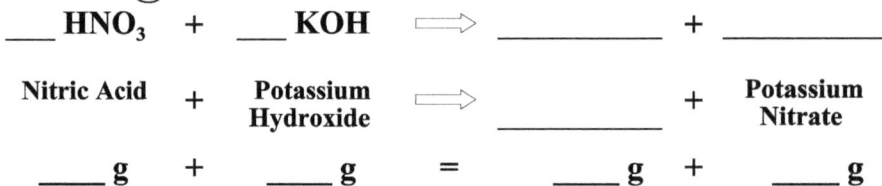

$(H)^+$ (O)(N)(O)(O)$^-$ $[K]^+$ $(O)(H)^-$ $\Rightarrow$

___ HNO_3 + ___ KOH $\Rightarrow$ _____ + _____

Nitric Acid + **Potassium Hydroxide** $\Rightarrow$ _____ + **Potassium Nitrate**

___ g + ___ g = ___ g + ___ g

Date: ___ / 5

Quiz 3.8: Reactions of Acids and Bases   Name:

---

5 Complete all four representations of this *neutralization* reaction between sulfuric acid and calcium hydroxide.

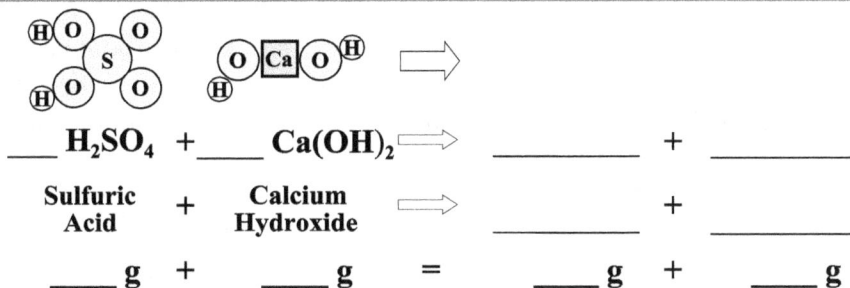

$(H)(O)$ (O) S (O) (O) (H) $(O)[Ca](O)(H)$ ⟹

___ H_2SO_4 +___ $Ca(OH)_2$ ⟹ _____ + _____

Sulfuric Acid + **Calcium Hydroxide** ⟹ _____ + _____

___ g + ___ g = ___ g + ___ g

Date: _____   / 5

---

6 When metal reacts with acid, hydrogen gas is produced, along with a salt. Complete all four representations of this reaction.

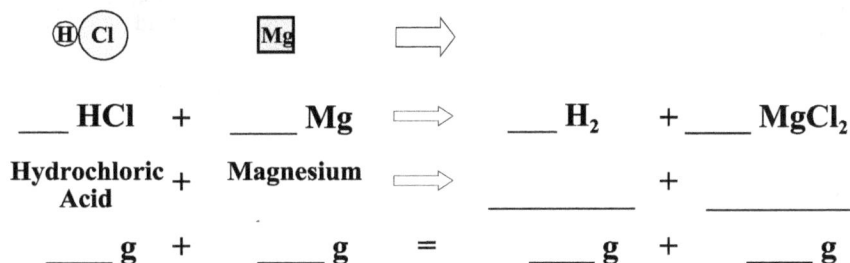

$(H)(Cl)$ $[Mg]$ ⟹

___ HCl + ____ Mg ⟹ ___ H_2 + ____ $MgCl_2$

Hydrochloric Acid + **Magnesium** ⟹ _____ + _____

___ g + ___ g = ___ g + ___ g

Date: _____   / 5

---

7 Complete all four representations of this *neutralization* reaction.

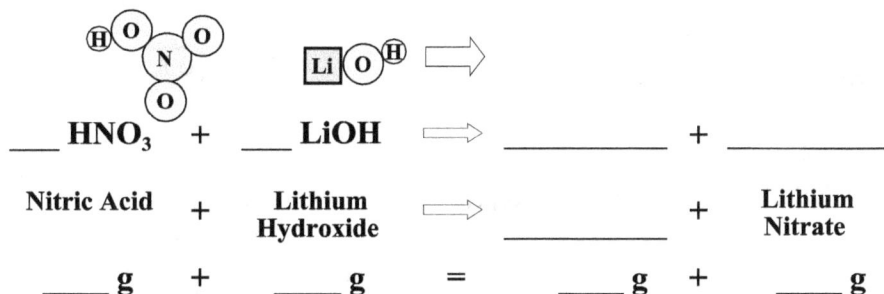

$(H)(O)$ N (O) (O) $[Li](O)(H)$ ⟹

___ HNO_3 + ___ LiOH ⟹ _____ + _____

Nitric Acid + **Lithium Hydroxide** ⟹ _____ + **Lithium Nitrate**

___ g + ___ g = ___ g + ___ g

Date: _____   / 5

---

8 Hydrobromic acid reacts with the metal aluminum to produce hydrogen gas plus a salt. Complete these representations. Count all the atoms!

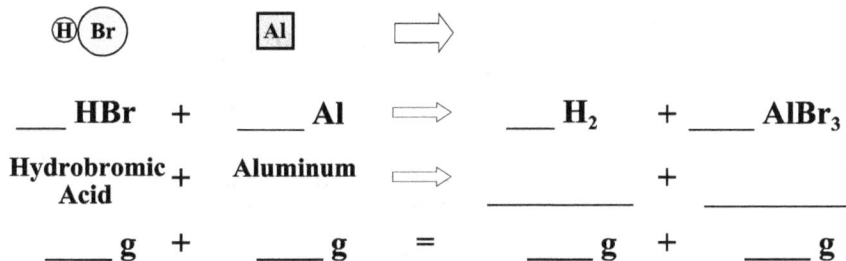

$(H)(Br)$ $[Al]$ ⟹

___ HBr + ____ Al ⟹ ___ H_2 + ____ $AlBr_3$

Hydrobromic Acid + **Aluminum** ⟹ _____ + _____

___ g + ___ g = ___ g + ___ g

Date: _____   / 5

Quiz 3.8: Reactions of Acids and Bases Name:

9 Carbon dioxide reacts with water to make carbonic acid. Complete all four representations of this synthesis reaction.

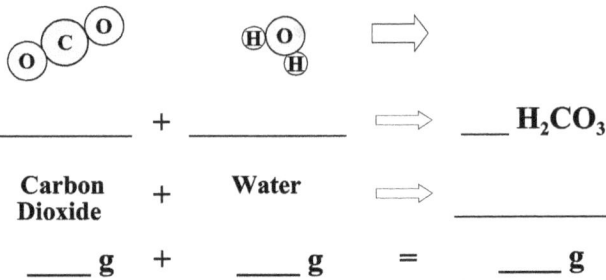

_____ + _____ ⇒ ___ H_2CO_3

Carbon Dioxide + **Water** ⇒ _____

___ g + ___ g = ___ g

Date: ____ / 5

10 Carbonic acid can decomposes into water and carbon dioxide. Complete all four representations of this decomposition reaction

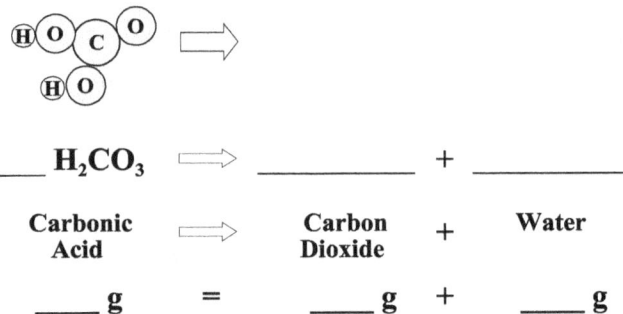

___ H_2CO_3 ⇒ _____ + _____

Carbonic Acid ⇒ **Carbon Dioxide** + **Water**

___ g = ___ g + ___ g

Date: ____ / 5

11 How many sodium hydroxide molecules must react to neutralize on molecule of carbonic acid? Complete all four representations.

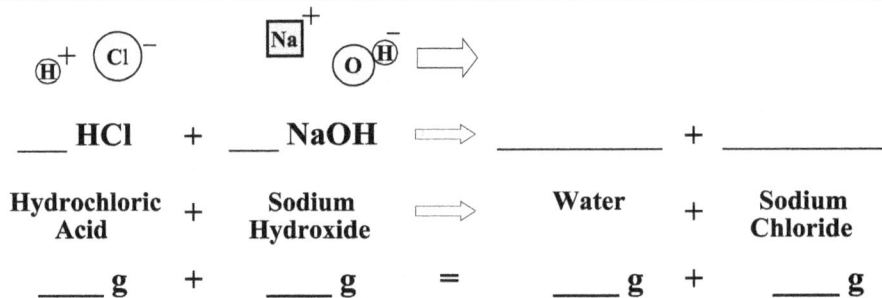

___ HCl + ___ NaOH ⇒ _____ + _____

Hydrochloric Acid + **Sodium Hydroxide** ⇒ **Water** + **Sodium Chloride**

___ g + ___ g = ___ g + ___ g

Date: ____ / 5

12 Acid causes sodium carbonate to decompose into simpler substances. What are the products? How many molecules of each?

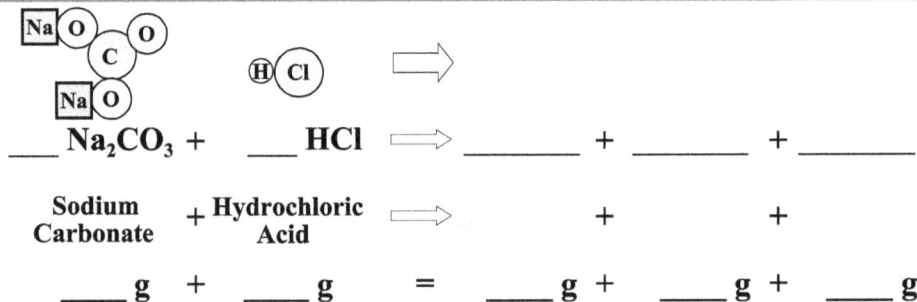

___ Na_2CO_3 + ___ HCl ⇒ _____ + _____ + _____

Sodium Carbonate + **Hydrochloric Acid** ⇒ _____ + _____ + _____

___ g + ___ g = ___ g + ___ g + ___ g

Date: ____ / 5

The Grade Ten Daily

Quiz 3.8: Reactions of Acids and Bases   Name:

---

13 Mark reactions **S**ynthesis, **D**ecomposition, **S**ingle Displacement, or **D**ouble Displacement.

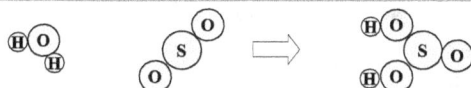

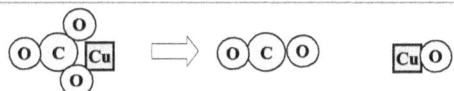

Date: _____   / 5

---

14 Mark reactions **S**ynthesis, **D**ecomposition, **S**ingle Displacement, or **D**ouble Displacement.

$$H_2SO_4 + Ca(OH)_2 \Longrightarrow CaSO_4 + 2\,H_2O$$

$$2\,H_2O_2 \Longrightarrow O_2 + 2\,H_2O$$

$$6\,HNO_3 + 2\,Fe \Longrightarrow 2\,Fe(NO_3)_3 + 3\,H_2$$

$$2\,NH_3 + H_2O + CO_2 \Longrightarrow (NH_4)_2CO_3$$

$$3\,CuCl_2 + 2\,Al \Longrightarrow 2\,AlCl_3 + 3\,Cu$$

Date: _____   / 5

---

15 Copper carbonate reacts with hydrochloric acid to make copper chloride, carbon dioxide, and water. Complete all four representations.

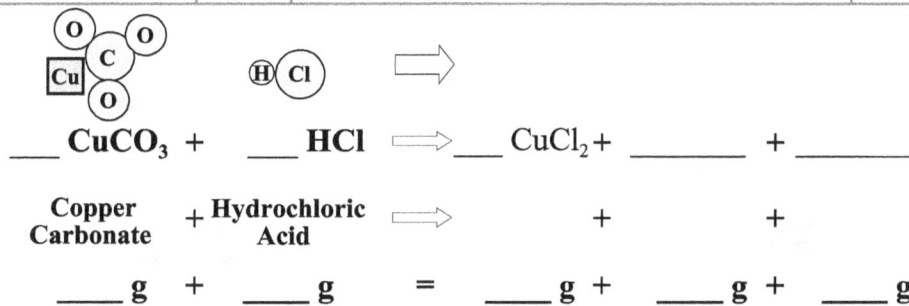

___ $CuCO_3$ + ___ $HCl \Longrightarrow$ ___ $CuCl_2$ + _____ + _____

Copper Carbonate + **Hydrochloric Acid** $\Longrightarrow$ _____ + _____ + _____

___ g + ___ g = ___ g + ___ g + ___ g

Date: _____   / 5

---

16 Complete all four representations of the reaction between copper chloride and aluminum.

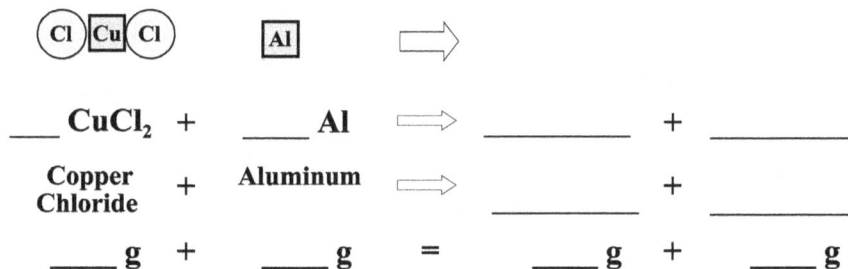

___ $CuCl_2$ + ___ $Al \Longrightarrow$ _____ + _____

Copper Chloride + **Aluminum** $\Longrightarrow$ _____ + _____

___ g + ___ g = ___ g + ___ g

Date: _____   / 5

Core Valence Radius Periodic Table

Core -Valence -Radius Periodic Table

The core charge is the sum of the positive protons in the nucleus plus the core electrons
The valence electrons are the outermost electrons
The radius is drawn to scale for each atom in this table
Electronegativity: the relative ability of an atom to grab and hold electrons from another atom

2.1							
(+1) H 1							(+2) He 2
1.0 (+1) Li 3	1.5 (+2) Be 4	2.0 (+3) B 5	2.5 (+4) C 6	3.0 (+5) N 7	3.5 (+6) O 8	4.0 (+7) F 9	(+8) Ne 10
0.9 (+1) Na 11	1.2 (+2) Mg 12	1.5 (+3) Al 13	1.8 (+4) Si 14	2.1 (+5) P 15	2.5 (+6) S 16	3.0 (+7) Cl 17	(+8) Ar 18
0.8 (+1) K 19	1.0 (+2) Ca 20	1.6 (+3) Ga 31	1.8 (+4) Ge 32	2.0 (+5) As 33	2.4 (+6) Se 34	2.8 (+7) Br 35	(+8) Kr 36

© Ross Lattner Publishing   82   www.rosslattner.com

Standard Periodic Table

The Standard Periodic Table

Legend:
- Atomic Number (top left)
- Electronegativity (top right)
- **Symbol**
- Name
- Mass

1	2	3	4	5	6	7	8	9	10	11	12	13	14	15	16	17	18
1 **H** Hydrogen 1.01 (2.1)																	2 **He** Helium 4.00 (–)
3 **Li** Lithium 6.9 (1.0)	4 **Be** Beryllium 9.01 (1.5)											5 **B** Boron 10.8 (2.0)	6 **C** Carbon 12.01 (2.5)	7 **N** Nitrogen 14.01 (3.0)	8 **O** Oxygen 16.00 (3.5)	9 **F** Fluorine 19.0 (4.0)	10 **Ne** Neon 20.2 (–)
11 **Na** Sodium 23.0 (0.9)	12 **Mg** Magnesium 24.3 (1.2)											13 **Al** Aluminum 27.0 (1.5)	14 **Si** Silicon 28.1 (1.8)	15 **P** Phosphorus 31.0 (2.1)	16 **S** Sulfur 32.1 (2.5)	17 **Cl** Chlorine 35.5 (3.0)	18 **Ar** Argon 39.9 (–)
19 **K** Potassium 39.1 (0.8)	20 **Ca** Calcium 40.1 (1.0)	21 **Sc** Scandium 45.0 (1.3)	22 **Ti** Titanium 47.9 (1.5)	23 **V** Vanadium 50.9 (1.6)	24 **Cr** Chromium 52.0 (1.6)	25 **Mn** Manganese 54.9 (1.5)	26 **Fe** Iron 55.8 (1.8)	27 **Co** Cobalt 58.9 (1.8)	28 **Ni** Nickel 58.7 (1.8)	29 **Cu** Copper 63.5 (1.9)	30 **Zn** Zinc 65.4 (1.6)	31 **Ga** Gallium 69.7 (1.6)	32 **Ge** Germanium 72.6 (1.8)	33 **As** Arsenic 74.9 (2.0)	34 **Se** Selenium 79.0 (2.4)	35 **Br** Bromine 79.9 (2.8)	36 **Kr** Krypton 83.8 (–)
37 **Rb** Rubidium 85.5 (0.8)	38 **Sr** Strontium 87.6 (1.0)	39 **Y** Yttrium 88.9 (1.2)	40 **Zr** Zirconium 91.2 (1.4)	41 **Nb** Niobium 92.9 (1.6)	42 **Mo** Molybdenum 95.9 (1.8)	43 **Tc** Technicium (98) (1.9)	44 **Ru** Ruthenium 101.1 (2.2)	45 **Rh** Rhodium 102.9 (2.2)	46 **Pd** Palladium 106.4 (2.2)	47 **Ag** Silver 107.9 (1.9)	48 **Cd** Cadmium 112.4 (1.7)	49 **In** Indium 114.8 (1.7)	50 **Sn** Tin 118.7 (1.8)	51 **Sb** Antimony 121.8 (1.9)	52 **Te** Tellurium 127.6 (2.1)	53 **I** Iodine 126.9 (2.5)	54 **Xe** Xenon 131.3 (–)
55 **Cs** Cesium 133 (0.7)	56 **Ba** Barium 137.3 (0.9)	71 **Lu** Lutetium 175	72 **Hf** Hafnium 178.5 (1.3)	73 **Ta** Tantalum 180.9 (1.5)	74 **W** Tungsten 183.8 (1.7)	75 **Re** Rhenium 186.2 (1.9)	76 **Os** Osmium 190.2 (2.2)	77 **Ir** Iridium 192.2 (2.2)	78 **Pt** Platinum 195.1 (2.2)	79 **Au** Gold 197.0 (2.4)	80 **Hg** Mercury 200.6 (1.9)	81 **Tl** Thallium 204.4 (1.8)	82 **Pb** Lead 207.2 (1.8)	83 **Bi** Bismuth 209.0 (1.9)	84 **Po** Polonium (209) (2.0)	85 **At** Astatine (210) (2.2)	86 **Rn** Radon (222) (–)
87 **Fr** Francium (223) (0.7)	88 **Ra** Radium (226) (0.9)	103 **Lr** Lawrencium (256)	104 **Unc**	105 **Unp**	106 **Unh**	107 **Uns**	108 **Uno**	109 **Une**									

Lanthanide Series

57 **La** Lanthanum 138.9 (1.1)	58 **Ce** Cerium 140.1 (1.1)	59 **Pr** Praseodymium 140.9 (1.1)	60 **Nd** Neodymium 144.2 (1.1)	61 **Pm** Promethium (145) (1.1)	62 **Sm** Samarium 150.4 (1.1)	63 **Eu** Europium 150.2 (1.1)	64 **Gd** Gadolinium 157.2 (1.1)	65 **Tb** Terbium 158.9 (1.1)	66 **Dy** Dysprosium 162.5 (1.1)	67 **Ho** Holmium 162.5 (1.1)	68 **Er** Erbium 167.3 (1.1)	69 **Tm** Thulium 168.9 (1.1)	70 **Yb** Ytterbium 173.0 (1.1)

Actinide Series

89 **Ac** Actinium (227) (1.1)	90 **Th** Thorium 232.0 (1.1)	91 **Pa** Protactinium (231) (1.1)	92 **U** Uranium 238.0 (1.5)	93 **Np** Neptunium (237) (1.7)	94 **Pu** Plutonium (244) (1.3)	95 **Am** Americium (243) (1.3)	96 **Cm** Curium (247) (1.3)	97 **Bk** Berkelium (247) (1.3)	98 **Cf** Californium (251) (1.3)	99 **Es** Einsteinium (254) (1.3)	100 **Fm** Fermium (253) (1.3)	101 **Md** Mendelevium (257) (1.3)	102 **No** Nobelium (255) (1.3)

Polyatomic Ions

Ion	Formula	Ion	Formula	Ion	Formula
Ammonium	NH_4^+	Nitrate	NO_3^-	Sulfate	SO_4^{2-}
Hydroxide	OH^-	Iodate	IO_3^-	Hydrogen Sulfate	HSO_4^-
Cyanide	CN^-	Nitrite	NO_2^-	Sulfite	SO_3^{2-}
Cyanate	OCN^-	Iodite	IO_2^-	Hydrogen Sulfite	HSO_3^-
Thiocyanate	SCN^-	Chlorate	ClO_3^-	Carbonate	CO_3^{2-}
Acetate	$C_2H_3O_2^-$	Bromate	BrO_3^-	Hydrogen Carbonate	HCO_3^-
Thiosulfate	$S_2O_3^{2-}$	Chlorite	ClO_2^-	Chromate	CrO_4^{2-}
		Bromite	BrO_2^-	Dichromate	$Cr_2O_7^{2-}$
		Hypochlorite	ClO^-	Phosphate	PO_4^{3-}
		Hypobromite	BrO^-	Hydrogen Phosphate	HPO_4^{2-}
		Perchlorate	ClO_4^-	Dihydrogen Phosphate	$H_2PO_4^-$
		Perbromate	BrO_4^-		
		Permanganate	MnO_4^-		

Appendix: Laboratory Safety

The Hazards **In this column is a list of lab safety issues that you will face in this course**	The Safe Way **Read this column to find out how to safely handle the laboratory problem.**
Eye Injury is possible from flying fragments of metal, glass or chemicals; from heat or flames; from caustic solutions such as acids or bases.	*Always wear safety glasses* in the laboratory. Never take your glasses off, even if you have finished your experiment. Other students may not have finished theirs. The safety glass symbol indicates exercises in which safety glasses *must* be worn.
Crowding, Pushing and Horseplay increase the likelihood of a serious injury.	*Attend to your work.* Stay at the station you were assigned, so that there is room to work safely. If your teacher finds that your behaviour is a safety hazard, he or she may remove you from the lab. There is no place for behaviours which place others at risk of injury. Not at school, not at home and not at work.
Disorganized and Dirty Working Conditions are a hazard wherever they are found.	*Keep Lab Area Clean.* Clean and put away unused equipment. Tell your teacher about chipped, cracked, damaged or broken equipment. Do not leave anything on the floor, the desktop, the sink, or the cupboards that is not supposed to be there.
Broken Glass happens even to careful scientists.	*Do Not Touch* broken glass with your hands. Tell your teacher. Use a broom to sweep the glass into a dustpan. Dispose of the broken glass in the special container provided. Do not leave it in the regular wastebasket: it could seriously injure a custodian.
Liquid Spills may consist of water, but they may also contain acids, bases, or toxic chemicals. You may not be able to tell the difference.	*Tell your teacher* about any spills immediately. Do not attempt to clean up without teacher instruction. Only if the teacher decides it's safe, use a cloth or paper towels to soak up excess liquid. Wipe the area clean with a damp cloth. Rinse the cloth frequently in fresh water. Wash your hands afterwards.
Solid Spills may consist of highly reactive chemicals. You may not know the specific hazards.	*Tell Your Teacher* about the spill, whether or not you caused it. Your teacher will instruct you on the safe way to handle the problem. In any case, the spill must be cleaned up promptly.

Appendix: Laboratory Safety

Open Flames are a frequent hazard. The Bunsen burner is the most likely safety hazard.	***Review Safe Handling of Bunsen Burner*** with your teacher. Be prepared to show how to light, operate and extinguish the burner at any time. Do not attempt to ignite pens, papers, rulers or other things. That kind of behaviour will certainly result in your being put out of the lab.
Fire. Any liquid solid or gaseous fuel burning where you do not want it to burn is a fire.	***Tell the teacher immediately!*** Do not attempt to extinguish the fire with your hands, books, paper towels etc. Do not panic. Move away from the hazard. ***Your teacher is the best judge of the appropriate course of action.***
Hot Metal or Glass cause more burns than any other hazard. There is usually no visible indication that they are hot. Glass in particular causes small, deep burns.	***Let Hot Objects Cool for 10 - 15 Minutes*** before handling. Place all hot objects on a heat resistant pad. You and your partner will know where they are. Approach hot objects cautiously. Touch them at the coolest point first (the base of the retort rod, the bottom of the Bunsen burner or hot plate, the thumb screw of the iron ring). Use dry, not damp, paper towels to handle hot objects.
Hot Liquids such as boiling water or hot oil spread and splash rapidly. They also cling to skin and clothes.	***Let Hot Liquids Cool for 10 - 15 Minutes*** before handling. Do not heat liquids in closed containers. Use hot plates rather than shaky retort rod assemblies. Do not heat more liquid than you need.
Obstructed Passageways prevent you from moving out the way of a spill or a fire.	***Stand at Your Lab Station.*** Do not bring chairs or stools over to sit down. Your chair will prevent others from moving away from a spill or a fire.
Long Hair or Loose Clothing is more likely to become involved in your equipment. It can cause spills and breakage, or catch fire.	***Tie Back Long Hair; Secure Loose Clothing.*** Outerwear in particular must be avoided in the lab situation. Jackets, sweat suits, hoods, etc are too large and awkward for the lab situation. They are also frequently made of materials that are flammable and can melt and stick to the skin in a fire. Avoid using laquer based hair sprays. A curly head of hair with hair spray can burn up completely in seconds.
Unauthorized Experiments can have unintended results.	***Stick to the plan.*** Read instructions very carefully the night before the lab. Ask questions. Do not try experiments "just to see what happens." The dangers are too great.

www.ingramcontent.com/pod-product-compliance
Lightning Source LLC
Chambersburg PA
CBHW081258110426

42743CB00045B/3309